NASA Conference Publication 2033

I0493829

Jet Aircraft Hydrocarbon Fuels Technology

John P. Longwell, Editor
Massachusetts Institute of Technology

A workshop held at
Lewis Research Center
Cleveland, Ohio
June 7-9, 1977

National Aeronautics and
Space Administration

Scientific and Technical
Information Office

1978

FOREWORD

A $2\frac{1}{2}$-day workshop was held at the NASA Lewis Research Center, Cleveland, Ohio, to study current jet fuel specifications and the need for future changes and trade-offs. The active participants in the workshop were representatives of NASA and other government agencies; airframe, engine, and petroleum industries; commercial airlines; and universities and consultants. Professor John P. Longwell of the Massachusetts Institute of Technology was the chairman of the workshop. Members of the Advanced Technology Section of the Lewis Research Center Airbreathing Engines Division organized the workshop and acted as the NASA liaisons within each of five working groups.

This report, prepared by Prof. Longwell, summarizes the findings, conclusions, and recommendations of the working groups. Specifications for an experimental, referee, broad-specification jet fuel are presented. These specifications were developed by Prof. Longwell from the recommendations of the workshop participants and through later consultations with NASA representatives and others.

Jack Grobman
NASA Lewis Research Center

CONTENTS

JET AIRCRAFT HYDROCARBON FUELS TECHNOLOGY

INTRODUCTION

The current jet fuel specifications were framed in the late 1940's and early 1950's and were written to take advantage of the abundance of high-quality middle distillates from petroleum. These specifications have served well and have survived with only minor modifications; however, it is highly probable that the availability of these naturally occurring high-quality fuels will diminish drastically during the remainder of this century. In preparation for such changes, the National Aeronautics and Space Administration and the Department of Defense have initiated programs aimed at assessing the suitability of fuels made from other materials, such as oil shale and coal, and are reexamining the fuel composition requirements for future aircraft. According to these studies, satisfactory jet fuels can be manufactured from these materials and also from tar and high-boiling-point petroleum fractions. To meet current specifications, however, major boiling range conversion and hydrogenation are required with an accompanying increase in cost and substantial energy consumption in the refining process. This situation calls for reexamining the trade-offs between fuel specifications and aircraft and engine design in order to determine the optimum combination for future aircraft. Such a program calls for contributions from engine and airframe researchers and designers, commercial and military operators, fuel manufacturers, and related research and government regulatory agencies.

For the past 2 years the NASA Ad Hoc Panel on Jet Engine Hydrocarbon Fuels, made up of individuals from these areas, had met several times. On completing its assignment, the panel recommended that a workshop on jet aircraft hydrocarbon fuels technology be held with an expanded group of participants and with the following purposes:

1. To obtain broad-based advice on appropriate specifications for referee fuels for research and development (R&D) use
2. To obtain advice on priority items for inclusion in the NASA-sponsored fuel R&D program

The first of these purposes is based on the conclusion of the ad hoc panel that a standardized fuel should be developed for use in R&D programs involving new combustion systems, engines, and aircraft and for testing in existing aircraft. Such a fuel will greatly simplify and accelerate the process of acquiring information on the problems, designs, and design trade-offs involved in using those fuels that could be efficiently manufactured in the context of the future supply situation.

A target time period of 1990 to 2000 was chosen. During this period it is believed that petroleum crudes will still furnish most of the hydrocarbon fuel liquids but that there will be sharp competition for the available supply of middle distillates. Higher-boiling-point petroleum fractions would be converted by "cracking" in order to supply the increased demand for fuels boiling in the kerosene boiling range. The cracked products are high in aromatics and require hydrogen treating to meet present jet fuel specifications. Boiling range conversion would be minimized if the final boiling point were increased or the flashpoint decreased. Energy consumption for hydrogenation would be minimized by allowing a larger aromatic content. These changes are accompanied by greater potential design and operating problems. The purpose of research and development relating to these problems is to reduce them to an acceptable level with the optimum compromise between aircraft and engine design and fuel manufacture and supply.

This report summarizes the findings and conclusions of this workshop. The workshop opened with an introductory session for all participants in which background papers were presented. The subjects and authors of these papers were as follows:

1. Introductory Remarks: Characteristics of the 1990-2000 Period. John P. Longwell, Massachusetts Institute of Technology
2. NASA Fuels Technology Program Overview. Jack Grobman, NASA Lewis Research Center
3. Forecast of Future Aviation Fuels. J. Morley English, University of California, Los Angeles
4. Fuel Users' Problems. Paul P. Campbell, United Airlines
5. Fuel Suppliers' Problems. William G. Dukek, Exxon Research and Engineering

The visual aid material employed in these presentations is attached as appendix A.

After an introductory session, the attendees were organized into five working groups:

1. Aviation Fuels Supply and Demand. Chairman, J. Morley English, UCLA
2. Combustor Research and Technology for Broad-Specification Fuels. Chairman, Donald W. Bahr, General Electric Co.
3. Fuel System Research and Technology for Broad-Specification Fuels. Chairman, Ivor Thomas, Boeing Commercial Airplane Co.
4. Fuel Thermal Stability Research and Technology. Chairman, Jack A. Bert, Chevron Research
5. Fuel Safety. Chairman, John H. Warren, Mobil Oil Sales and Supply Corp.

A list of working group participants is attached. Brief reports were prepared by these working groups, and the reports were presented for discussion to all attendees at the final session of the workshop. The groups' conclusions and recommendations are attached as appendix B.

To aid in focusing working group discussions, tentative specifications for broadened-composition jet fuels were distributed to the participants. The recommended experimental referee fuel specification was developed from the comments and recommendations of the attendees during the final session and by further consultation following the meeting.

2

EXPERIMENTAL REFEREE BROAD-SPECIFICATION FUEL

The workshop's recommended broad-specification referee fuel to be used as a base-case fuel in the programs aimed at developing new engines and fuel systems is described in table I. This fuel is designed particularly for combustion system research by control of fuel composition. While studies of fuel stability, low-temperature handling, or fire safety might require special control and variation of the relevant characteristics, an attempt was made to choose all properties to be mutually consistent and to be within a reasonable range. The specifications are written to avoid redundancy, to be reasonably easy to supply, and to be sufficiently reproducible from the viewpoint of combustion characteristics.

FUEL COMPOSITION

Projections of future sources of supply and future uses of liquid hydrocarbons indicate that an increased fraction of petroleum will go into middle-distillate products (jet fuel, diesel fuel, No. 2 heating oil, kerosene, and petrochemical feed). Since straight distilled product is insufficient for the demand, middle-distillate fractions will be produced by cracking of higher-boiling-point materials. The higher aromatic content of cracked products, increasing competition for paraffinic materials for petrochemical and diesel fuel use, and growth in kerosene-type jet fuel use by both commerical and military aviation will force the use of energy-intensive hydrogenation processes if current specifications are to be maintained. Examination of the composition of cracked streams suggests that a maximum aromatics content of 35 percent should be adequate to cover expected aromatics increases as long as coal liquids are not used in significant quantity.

It was generally agreed that hydrogen content is a better measure of combustion properties than the currently used aromatics content, and the specification was based on this measure. The mean value of 12.8 was chosen to correspond to 35-percent aromatics. It should be noted, however, that the hydrogen content and the aromatics content are not uniquely related. In current specifications, a maximum is set on the aromatics content, but for this referee fuel, both a lower and an upper limit were set, in effect, by the lower and upper limits on the hydrogen content.

It was also recommended that detailed analyses of major batches of referee fuels be made and reported. It is now practical to obtain such information; and, if variations in combustion performance between fuel batches are observed, study of such analyses will offer a means of explanation and control. Specifications of other composition variables such as sulfur content were retained as given in American Society of Testing and Materials procedure ASTM D1655-77. A nitrogen specification was suggested but was sub-

3

TABLE I. - PROPOSED SPECIFICATIONS FOR EXPERIMENTAL

REFEREE BROAD-SPECIFICATION (ERBS)

AVIATION TURBINE FUEL

Specifications	ERBS jet fuel value	Proposed test method
Composition:		
Hydrogen, wt%	12.8±0.2	NMR
Aromatics, vol%	Report	ASTM D1319
Sulfur, mercaptan, wt%	0.003, max.	ASTM D1219
Sulfur, total, wt%	0.3, max.	ASTM D1266
Nitrogen, total, wt%	Report	Kjeldahl
Naphthalenes, vol%	Report	ASTM D1840
Hydrocarbon compositional analysis	Report	GCMS
Volatility:		
Distillation temperature, °F:		ASTM D2892
Initial boiling point	Report	↓
10 Percent	400, max.	
50 Percent	Report	
90 Percent	500, min.	
Final boiling point	Report	
Residue, percent	Report	
Loss, percent	Report	
Flashpoint, °F	110±10	ASTM D56
Gravity, API (60° F)	Report	ASTM D287
Gravity, specific (60/60° F)	Report	ASTM D1298
Fluidity:		
Freezing point, °F	-20, max.	ASTM D2386
Viscosity, at -10° F, cS	12, max.	ASTM D445
Combustion:		
Net heat of combustion, Btu/lb	Report	ASTM D2382
Thermal stability:		
JFTOT, breakpoint temperature, °F (TDR, 13; and ΔP, 25 mm)	460 min.	ASTM D3241

sequently judged to be unnecessary since attainment of the specified fuel stability is expected to require a nitrogen content below the level at which it will contribute significantly to nitric oxide formation.

It is probable that parametric variation of hydrogen content both above and below the recommended mean value of 12.8 will be required for combustor research and development and for testing in full-scale engines. A desirable technique would be to acquire two standardized batches of fuel: one with a high hydrogen content, and the other with significantly lower hydrogen content. Both materials should meet the other specifications for the experimental, referee, broad-specification (ERBS) fuel. Fuels of intermediate properties could then be easily prepared by blending with the base-case fuels. The high-hydrogen-content fuel could be readily prepared by distilling a high-quality crude. The low-hydrogen-content fuel will be more difficult to acquire and will probably require a search for special blending stocks from extraction processes or other special sources.

VOLATILITY, FLUIDITY, AND COMBUSTION

Fuel volatility is primarily set by flashpoint at the low end of the boiling range and by freezing point at the high end. Decreasing the flashpoint is probably an effective and efficient way of extending the fractions of petroleum that can be used in jet fuel. Reduction of the present flashpoint specification to perhaps 80° F could be acceptable from a safety viewpoint (summary of working group V, appendix B). However, it was decided that the 100° F minimum should be retained for the purposes for which this special experimental fuel was intended. The 400° F maximum specification for the 10-percent distillation point insures a reproducible amount of the lighter fractions and sets the flashpoint in the range 100° to 120° F. To assure inclusion of higher-boiling-point fractions, the 90-percent point was set at 500° F minimum. Additional intermediate-boiling-range specifications were not believed to be necessary for a normal boiling range distribution between the 10- and 90-percent points. The final boiling point is set by the freezing point specification.

Working Group III, on fuel system R&T, concluded that a -20° F freezing point represented a practical maximum since the cost and complexity of dealing with higher freezing points, such as 0° F, increase rapidly. While the freezing point per se does not interact strongly with combustion, it does set the maximum final boiling point (at around 600° F for -20° F freezing point). This is lower than the 650° F typical of diesel fuel and No. 2 heating oil. Parametric combustion and fuel system studies of fuels with a 650° F final boiling point were strongly recommended. For such fuels the 90-percent point would also be increased. This might best be accomplished by blending additional high-boiling-point fractions into the standard ERBS fuel. These high-boiling-point fractions should be chosen so as to change composition and other properties by a minimum amount. Reduction in 90-percent point might best be accomplished by redistilling the standard ERBS fuel. Freezing point as well as combustion properties would be changed. Realistic independent variation of freezing point would probably require fuels that meet ERBS specifications but are manufactured from different crude-oil sources.

Specific gravity and heat of combustion were not specified since they will be set within quite narrow limits by the specification of boiling range and the hydrogen content.

THERMAL STABILITY

Meeting requirements for thermal stability will be increasingly difficult as the aromatics content and the final boiling point increase and as cracked stocks and high-nitrogen-content sources such as shale oil are used.

The ASTM Jet Fuel Thermal Oxidation Test (JFTOT) was chosen for use with the lower breakpoint temperature of 460^0 F used in ASTM D1655-77, rather than with the 500^0 F breakpoint also used in ASTM D1655-77. This choice was not discussed in detail at the workshop, but subsequent consideration led to recommending the lower value on the basis that it would ease the problem of acquiring test fuels manufactured from cracked stocks. The lower breakpoint temperature would also accelerate the identification and solution of fuel-stability-related problems in research and development work. This specification will require hydrotreating to reduce the nitrogen and olefin contents and will also require care in fuel system design to avoid overheating. Study of fuel stability and of fuel system design for reduced stress on the fuel is an important component of the recommended program and will require study of a wide variety of fuels apart from the base-case referee fuel.

SUMMARY OF WORKING GROUP CONCLUSIONS AND RECOMMENDATIONS

This summary is based on reports prepared by the working groups, which are presented in appendix B.

6

WORKING GROUP I - AVIATION FUELS SUPPLY AND DEMAND

General Findings and Conclusions

1. The demand and supply outlook for aviation fuels in the 1990's reinforces the proposal that new aircraft should be capable of operating on a wider-cut fuel than jet kerosene. A broad-specification, petroleum-based fuel similar to No. 2 diesel provides refiners with the maximum flexibility for meeting jet fuel demand with the least costly processing sequence.

2. Such a fuel can be introduced into the world and U.S. market in several ways. Finding the optimum way requires study of current aircraft and ground distribution systems and trade-offs between fuel manufacturing problems and airline penalties.

3. Since it is undesirable to have more than one fuel supply system at airports and since retrofitting programs are also undesirable, acquiring the data base and studies needed to resolve this dilemma should be an important part of the fuels R&D program.

Specific Recommendations for Studies

1. Current aircraft. - To what extent and at what cost can remaining current-model aircraft be retrofitted, beginning about 1990, to use broad-specification fuels?

2. Ground distribution systems. - What is the investment/operating cost comparison between a system-wide introduction of a new jet fuel in the United States in 1990 and the introduction of a separate distribution system for domestic use of the broad-specification fuel while maintaining Jet A for international and other special operations?

3. Cost/penalty trade-offs. - How is the relative cost advantage of a broad-specification fuel (compared with Jet A fuel) best assessed against the penalty of operating new aircraft or retrofitted aircraft adapted for such a fuel?

WORKING GROUP II - COMBUSTION RESEARCH AND TECHNOLOGY FOR

BROAD-SPECIFICATION FUELS

General Findings and Conclusions

1. It is likely that, with current- and advanced-technology combustor designs, broad-specification fuels will yield higher levels of carbon monoxide, hydrocarbons, oxides of nitrogen, and smoke emissions than present-day jet fuels. Appropriate adjustments of the Environmental Protection Agency emissions standards to accommodate

these new fuels will probably be needed. The EPA should be continually provided with pertinent data so that appropriate and timely adjustments to standards can be defined.

2. Viscosity for any fuel must be limited to a maximum value of about 15 centistokes in order to obtain acceptable ignitiion and altitude relight performance. The broadened fuel specifications are near this maximum viscosity at -10° F, and this will limit the minimum fuel temperature at the combustor inlet. Fuel heaters will be needed on cold days. Also, special provisions will be necessary for satisfactory ignition at altitude relight operating conditions.

3. Additional data are needed on the basic combustion and atomization characteristics of low-hydrogen-content fuels. Furthermore, to characterize these fuels, more precise methods of measuring fuel hydrogen content are needed, to a capability within ±0.02 percent.

4. The possibility of defining a highly standardized and precisely characterized fuel blend for use in developing combustors intended for low-hydrogen-content fuels should be considered. Preferably, this blend should consist of known amounts of prescribed hydrocarbon constituents.

5. With the lower-hydrogen-content, broad-specification fuels, higher smoke levels and associated higher metal temperatures may be problems in current combustors. Increasing combustor dome airflow or liner cooling flows to reduce smoke and wall temperatures may adversely affect altitude relight capability or exit temperature profiles. On the other hand, advanced technology combustors with staged or variable geometry may minimize wall temperature problems without introducing other operational risks.

Specific Recommendations for Studies

1. Initiate or extend programs to provide more complete and better fundamental data on combustion and atomization properties of low-hydrogen-content fuels, conduct combustor rig tests with these fuels, and extend the data to engine testing.

2. Define and develop methods of measuring the properties and chemical constituents of fuels, especially the hydrogen content.

3. Initiate programs to define and develop technology for obtaining acceptable ignition and altitude relight capabilities, using the broad-specificiation fuels, with acceptable smoke levels. Specific approaches to be investigated include
    a. Alternative ignition methods
    b. Added fuel control system capabilities
    c. Alternative fuel injection methods
    d. Fuel staging methods

4. Initiate programs to define and develop technology for maintaining acceptable

dome and liner metal temperatures, using the broad-specification fuels, without relying on large increases in cooling airflows. Specific approaches to be investigated include

    a. Thermal barrier coatings

    b. Added cooling provisions

5. Conduct additional tests of existing staged combustors with the broad-specification fuels; develop, as required, the design improvements needed to accommodate the use of these fuels; and include the use of the new fuels in all future advanced technology combustor development programs.

WORKING GROUP III - FUEL SYSTEM RESEARCH AND TECHNOLOGY FOR

BROAD-SPECIFICATION FUELS

General Findings and Conclusions

1. From the viewpoint of fuel system technology the most important trend of fuel specification limits in the next 25 years will be holding the initial boiling point near its present limit but letting the final boiling point increase. This will primarily affect the freezing point.

2. Because of the projected rapid increase in operating cost as freezing point exceeds -20° F, it was considered unlikely that higher freezing points would be a reality; however, work with fuels up to 0° F freezing point is recommended to allow better identification of the optimum. It was considered that NASA's program in this area was satisfactory.

3. The problems introduced by increased aromatics content on water solubility, cleanliness, viscosity, and material compatibility should also be studied.

Specific Recommendations for Studies

1. Freezing point research. - The following recommendations were made for freezing-point research:

    a. A follow-on study to the recently completed Boeing program (NASA CR-135198) should be conducted to refine and detail heating systems that will permit the use of high-freezing-point fuels. The follow-on should include evaluation of the number of flights versus the minimum freezing point and the cost of operation versus the freezing point.

    b. Some effort should be made to support the business jet sector of the industry, who have little or no say in the fuel properties but are developing very long-range business jets.

c. Fuel freezing behavior in a fuel tank should be studied to determine if two-phase flow can be tolerated. Results should be aimed at refinery and laboratory testing of flowability and pumpability.

d. Full-scale testing on a ground fuel system simulator should be followed by a flight test (particularly if two-phase flow is shown to be feasible in the small-scale test).

e. A user "quick test" taking less than 10 minutes would be useful to the operators. It could help the operator avoid diversions or optimize flight plans against the actual freezing point rather than specification values. Some ideas for quick tests were mirror-dewpoint apparatus with a thermal electric cooler, a cooled bar inserted in the fuel to measure wax buildup, and analysis of normal paraffin content by molecular sieve absorption.

f. An analysis of in-flight temperatures compiled by the International Air Transport Association is desirable.

2. Aromatics. - The compatibility of elastomeric materials as a function of aromatic content, up to 30 or 40 percent, should be investigated. A correlation with hydrogen content or some other property would also be desirable in view of the range of aromatic compounds that might be present in a high-freezing-point fuel. One panel member suggested that any compatibility testing be conducted with a specific aromatic content defined for each portion of the distillation range of the test fuel. The work should include documentation of current materials used in commercial aircraft.

3. Water solubility and cleanliness. - The trends of these problems must be looked at, particularly with respect to aromatic content.

4. Viscosity. - This is not a problem in the fuel system unless the flow becomes non-Newtonian at very low temperatures, which will be examined as part of the freezing point program. However, it was noted that viscosity must be examined carefully with respect to engine starting.

WORKING GROUP IV - FUEL THERMAL STABILITY RESEARCH AND TECHNOLOGY

General Findings and Conclusions

1. The deposits related to fuel thermal stability are of significant concern with today's fuels and engines.

2. Decreasing the hydrogen content and increasing the final boiling point both tend to increase the problem.

3. Thermal stability is dependent on trace materials, so a generalized referee fuel will not adequately represent the potential for future fuel stability problems; however, it is expected that thermal stability of the contemplated broad-specification referee fuel will

range from a breakpoint of 400° F to less than 500° F.

4. Fuel system modifications to reduce thermal stress on the fuel can reduce deposits.

Specific Recommendations for Studies

1. There is a real need for a relationship between laboratory thermal stability test results and full-scale engine deposit results. Therefore, a research program that uses a fuel system simulator capable of covering all flight conditions is suggested in order to identify the sensitivity of the system to deposit formation. Results would aid in the design for minimum deposit formation and in predicting the effect of fuel quality. In essence, the simulator should be sized to be representative of a full-scale engine. To simplify design, requirements for takeoff capability might be compromised relative to air and fuel flow rates because nozzle fouling is minimal under takeoff conditions.

2. There should be continuing basic research on the controlling processes in deposit formation, such as surface materials and fluid mechanics.

3. Fundamental laboratory studies should be extended to include the investigation of the chemical composition of the fuel so as to identify the chemical species that are deleterious to thermal stability. In particular, constituents introduced by using fuel fractions heavier than those used in current kerosene-type jet fuels should be identified.

4. Deoxygenation and fuel additives for control of deposit formation should be investigated.

5. Deposits found in operational aircraft should be analyzed.

WORKING GROUP V - FUEL SAFETY

General Findings and Conclusions

1. The group saw no reason to invalidate the 1975 Coordinating Research Council Fuel Safety Report.

2. Some reduction of the fuel flashpoint below the present 100° F minimum is feasible from a safety standpoint. A level as low as 80° to 90° F was considered acceptable.

3. The wide variance of regulations by different municipalities, states, and countries will have a significant effect on the acceptability of reduced-flashpoint jet fuel. The marketing practice of dual branding of kerosene for ground and aviation use will strongly influence the acceptance of a lower-flashpoint product since flashpoints below 100° F are unsafe for home use.

Specific Recommendations for Studies

No specific activities are recommended for NASA at this time.

CONCLUDING REMARKS

The results of the Jet Aircraft Hydrocarbon Fuels Technology Workshop can be summarized in terms of the two purposes of the workshop:

1. To obtain broad-based advice on appropriate specifications for referee fuels for research and development use
2. To obtain advice on priority items for inclusion in the NASA-sponsored fuel research and development program

The experimental referee fuel specification is presented in this report. It was developed from the recommendations of the workshop attendees and from postmeeting discussions. The broad-specification fuel is designed particularly for combustion system research, but the properties were chosen to be consistent and practical, and they are also applicable to proposed fuel system research. The most significant changes in the proposed referee fuel are a lower and closely specified hydrogen content and a higher final boiling point and freezing point. It is hoped that petroleum refineries can produce such a referee, broad-specification fuel in sufficient quantities for experimental use and testing as a standardized fuel by the airframe and engine manufacturers, commercial and military users, and research organizations including NASA.

The workshop participants divided into five working groups to concentrate on particular areas of fuel research and technology. This report presents the findings, conclusions, and recommendations of the working groups in appendix B and in brief summary in the body of the report. Several significant recommendations were made by the groups in accordance with the second aim of the workshop.

Predictive studies of fuel supply and demand are essential. Continuation of present studies should include more extensive trade-off and cost-effectiveness predictions that relate the overall fuel refinery, ground distribution, and air transportation systems.

Combustion research on current- and advanced-technology designs must include performance with the broad-specification, referee fuel. Most likely, combustor problem areas with the new fuels will be in liner temperature control, because of increased flame radiation, ignition, and altitude relight capability, and in emission reduction.

The most important recommendation for aircraft fuel system studies with the broad-specification fuels is the extension of research to large-scale fuel systems, manifolds, and injector simulators. These would relate laboratory studies to actual performance in aircraft systems. The principal applications of the practical simulation studies would be to extend the recent studies on high-freezing-point fuel pumpability and behavior and to correlate laboratory thermal stability results to full-scale fuel deposit predictions.

APPENDIX A

FIGURES FROM INTRODUCTORY GENERAL PRESENTATIONS

INTRODUCTORY REMARKS: CHARACTERISTICS OF THE 1990-2000 PERIOD

John P. Longwell,

Massachusetts Institute of Technology

CHARACTERISTICS OF 1990 - 2000 PERIOD

- PRONOUNCED WORLDWIDE SHORTAGE OF PETROLEUM

- SMALL BUT GROWING PRODUCTION OF COAL AND SHALE LIQUIDS

- COMPETITION FOR PARAFFINIC MID-DISTILLATES

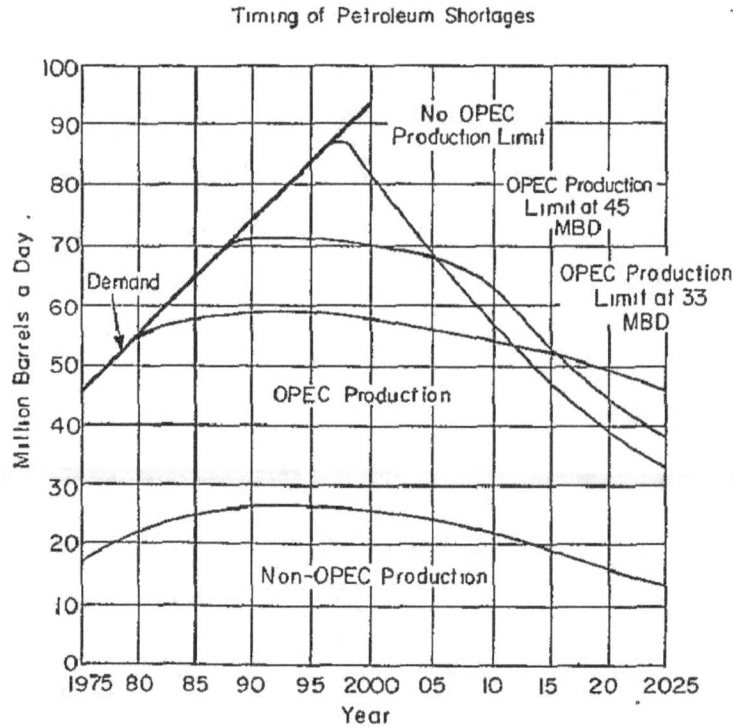

Timing of Petroleum Shortages

14

MAJOR LONG RANGE TRENDS IN PARAFFINIC FUEL DEMAND

INCREASED USE OF PARAFFINIC FUELS

- DIESEL FUEL
- KEROSENE TYPE JET FUEL
- PETROCHEMICAL FUEL

DECREASED DEMAND FOR AROMATIC FUELS

- FUEL OIL

STABLE OR SOMEWHAT DECREASED DEMAND FOR
- MOTOR GASOLINE
- HOME HEATING OIL

ESTIMATED SHIFTS IN PRODUCT DEMAND

	YEAR	
	1975	2000
MID-DISTILLATES/TOTAL LIQUID FUELS	.27	.4
MID-DISTILLATES/GASOLINE	.60	.95
JET + DIESEL/GASOLINE	.32	.65
JET/GASOLINE	.17	.27

PURPOSE OF R + D PROGRAM

- TO ACQUIRE THE INFORMATION NEEDED FOR ASSESSING THE TRADE-OFF

  BETWEEN FUELS MANUFACTURING EFFICIENCY AND COST, FUTURE AIRCRAFT

  AND ENGINE PERFORMANCE, AND OPERATIONAL PROBLEMS.

- TO CARRY OUT R + D PROGRAMS LEADING TO THE CAPABILITY OF USING

  BROAD SPECIFICATION FUELS IN NEW AIRCRAFT AND ENGINES.

A FUELS TECHNOLOGY PROGRAM OVERVIEW

Jack Grobman

NASA Lewis Research Center

EFFICIENT UTILIZATION OF FOSSIL FUELS FOR AVIATION

INVESTIGATE ALTERNATE SOURCES OF AIRCRAFT FUELS
  SHALE OIL
  COAL SYNCRUDE

MINIMIZE REFINERY ENERGY CONSUMPTION & REDUCE FUEL COST
  RELAX AIRCRAFT FUEL SPECIFICATIONS
  EVOLVE COMPONENT TECHNOLOGY TO PERMIT USE OF "BROAD-
    SPEC" AIRCRAFT FUEL

MINIMIZE AIRCRAFT FUEL USAGE
  REDUCE ENGINE SPECIFIC FUEL CONSUMPTION
  REDUCE ENGINE WEIGHT
  REDUCE AIRCRAFT WEIGHT   CS-77-590
  IMPROVE AIRCRAFT AERODYNAMICS

INTERAGENCY COORDINATION

- INTEGRATED USAF/NASA AIRCRAFT TURBINE FUELS TECHNOLOGY
  PROGRAM

- DOD SYNTHETIC FUELS R&D COORDINATION GROUP

  - NAVY, ARMY, AIR FORCE, ERDA, NASA

- NASA AD HOC PANEL ON JET ENGINE HYDROCARBON FUELS

  - AIR FORCE, NAVY, ERDA, FAA, P&W, GE, EXXON, TWA,
    UNITED, TEXACO

- NASA WORKSHOP ON JET AIRCRAFT HYDROCARBON FUELS
  TECHNOLOGY - JUNE 7-9, 1977

NASA FUELS TECHNOLOGY PROGRAM

OBJECTIVE
   EVALUATE THE POTENTIAL CHARACTERISTICS OF FUTURE JET
      AIRCRAFT FUELS, DETERMINE EFFECTS ON ENGINE COMPONENTS
      & EVOLVE COMPONENT TECHNOLOGY IF NEEDED

APPROACH
   IDENTIFY DEGREE TO WHICH FUEL SPECIFICATIONS MAY BE RELAXED
   DETERMINE EFFECTS OF RELAXING FUEL SPECIFICATIONS ON DESIGN
      OF:
         COMBUSTORS
         TURBINES
         FUEL TANKS
         FUEL SYSTEM
         MATERIALS
   PERFORM ENGINE DEMONSTRATION TESTS WITH CANDIDATE
      ALTERNATE FUELS

CS-77683

TARGET: TO IDENTIFY THE PROBABLE PROPERTIES OF FUTURE ALTERNATIVE AVIATION TURBINE
          FUELS REFINED FROM EITHER PETROLEUM, SHALE OIL OR COAL SYNCRUDES

PRINCIPAL PROGRAM ELEMENTS

IN-HOUSE

o   LABORATORY SYNTHESIS OF JET FUELS FROM SHALE OIL AND COAL SYNCRUDES (NASA)

o   LABORATORY SYNFUEL CHARACTERIZATION STUDIES (AF & NASA)

   -   COMBUSTION PROPERTIES, E.G., AROMATIC CONTENT

   -   THERMAL STABILITY

   -   FREEZING POINT

CONTRACT/GRANT

o   STUDY GRANT, "FORECAST OF FUTURE AVIATION FUELS", U. C. L. A. (NASA)

o   REFINERY ENERGY OPTIMIZATION STUDY, GORDIAN ASSOC.   (NASA)

o   LABORATORY SYNFUEL PROCESSING STUDIES (AF & NASA)

   -   EXXON

   -   ATLANTIC RICHFIELD

o   STUDY GRANT, "STABILITY OF NITROGEN-CONTAINING TURBINE FUELS," COLORADO SCHOOL
    OF MINES (NASA)

HYDROGEN CONSUMPTION IN PROCESSING

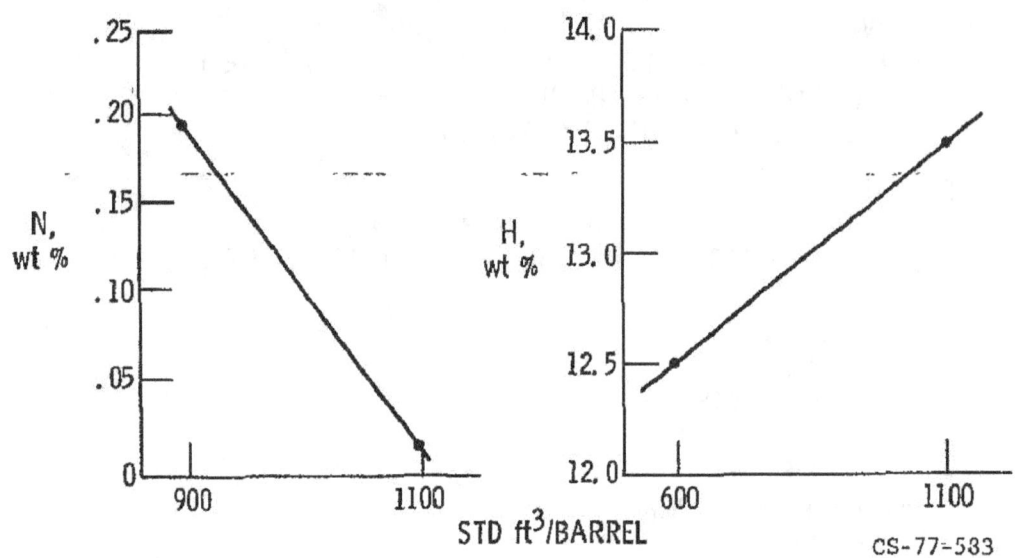

CS-77-583

VARIATION OF BREAKPOINT TEMP. WITH NITROGEN LEVEL

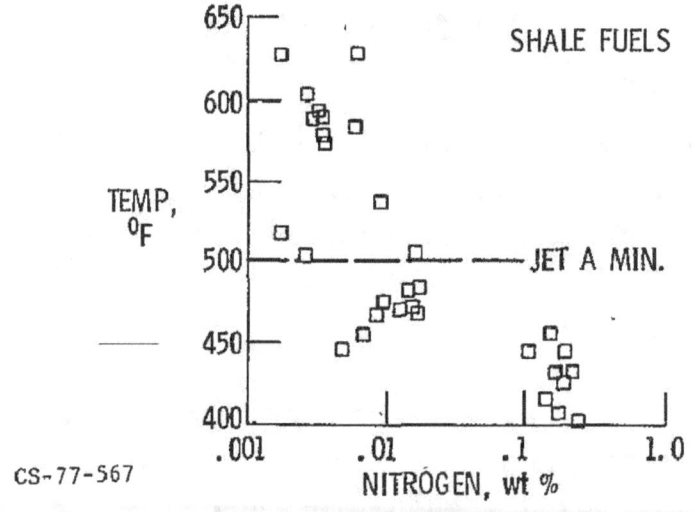

CS-77-567

18

TARGET: TO DETERMINE THE EFFECTS OF RELAXING FUEL SPECIFICATIONS ON COMBUSTOR
PERFORMANCE, EMISSIONS, AND DURABILITY AND TO EVOLVE AND EVALUATE COMBUSTOR
TECHNOLOGY FOR BROAD SPEC. FUELS

PRINCIPAL PROGRAM ELEMENTS

IN-HOUSE

o EXPERIMENTAL EVALUATION OF CONVENTIONAL COMBUSTORS WITH BROAD SPEC. FUELS (NASA & AF)

o EVOLUTION AND EVALUATION OF CONCEPTUAL COMBUSTORS FOR BROAD SPEC. FUELS (NASA)

o EXPERIMENTAL MEASUREMENT OF FLAME RADIATION FOR VARYING FUEL PROPERTIES (NASA)

CONTRACT/GRANT

o EXPERIMENTAL EVALUATION OF LOW-POLLUTANT COMBUSTORS WITH BROAD SPEC. FUELS: GE, P&W (NASA)

o COMBUSTOR DESIGN STUDY - BROAD SPEC. FUELS (NASA)

o STUDY OF EFFECT OF FUEL PROPERTIES ON SOOT FORMATION AND OXIDATION, M.I.T. (NASA)

o EVOLUTION AND EVALUATION OF COMBUSTOR TECHNOLOGY FOR BROAD SPEC. FUELS (NASA)

o EVALUATION OF COMBUSTORS FOR AF ENGINES WITH BROAD SPEC. FUELS (AF)

o EVALUATION OF STOICHIOMETRIC COMBUSTOR WITH BROAD SPEC. FUELS (AF)

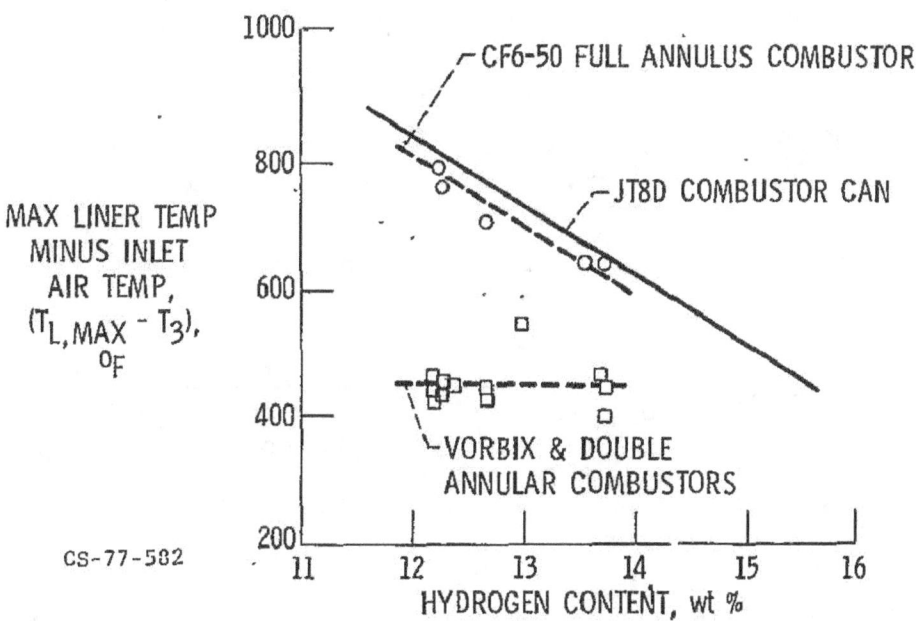

EFFECT OF HYDROGEN CONTENT OF FUEL ON
COMBUSTOR LINER SURFACE TEMPERATURE

19

TARGET: TO DETERMINE THE EFFECTS OF RELAXING FUEL SPECIFICATION ON FUEL SYSTEM
PERFORMANCE AND DURABILITY AND ON ENGINE MATERIALS, SUCH AS FUEL SYSTEM
ELASTOMERS AND HOT SECTION ALLOYS AND COATINGS, AND TO EVOLVE AND EVALUATE
TECHNOLOGY FOR BROAD SPEC. FUELS

PRINCIPAL PROGRAM ELEMENTS

IN-HOUSE

o   EVALUATION OF THERMAL BARRIER COATINGS FOR COMBUSTORS   (NASA)

o   ACCELERATED HOT CORROSION-TURBINE MATERIAL TESTS WITH ALTERNATIVE FUELS (NASA)

o   MATERIALS/FUEL COMPATIBILITY STUDY (AF)

CONTRACT

o   FUEL SYSTEM DESIGN STUDY FOR HIGH FREEZING POINT FUELS, BOEING   (NASA)

o   EXPERIMENTAL STUDY OF PUMPABILITY IN LOW TEMPERATURE FUEL SYSTEMS (NASA)

o   EXPERIMENTAL EVALUATION OF EFFECT OF FUEL PROPERTIES ON ELASTOMERS, JPL   (NASA)

o   AF ENGINE/AIRFRAME/FUEL TRADEOFF STUDY (AF)

FUEL TANK TEMPERATURES FOR 5000 n. mi. FLIGHT WITH HEATING

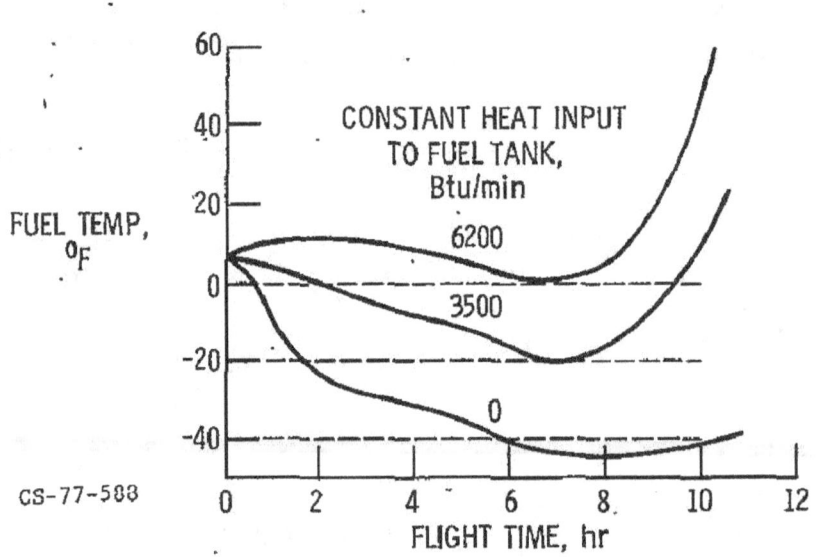

CS-77-588

ASSESSMENT OF POTENTIAL SOLUTIONS TO
JET FUEL PROBLEMS

SOLUTION	ADVANTAGES	DISADVANTAGES
PRODUCE SPECIFICATION JET FUEL	OPTIMIZED FUEL PROPERTIES AIRCRAFT/ENGINE RETROFIT NOT REQUIRED	INCREASED REFINERY ENERGY CONSUMPTION INCREASED FUEL COST
RELAX JET FUEL SPECIFICATION	CONSERVATION OF ENERGY REDUCED FUEL COST	MORE COMPLEX COMPONENT TECH REQUIRED ADVERSE EFFECT ON ENGINE LIFE

CS-77-581

RESEARCH & DEVELOPMENT REQUIRED TO
USE ALTERNATE FUELS

COMBUSTOR TECHNOLOGY
   HIGHER AROMATICS CONTENT
   LOWER VOLATILITY

FUEL SYSTEM TECHNOLOGY
   HIGHER FREEZING POINT
   LOWER THERMAL STABILITY

FUNDAMENTAL DATA RELATING THERMAL STABILITY WITH FUEL
   COMPOSITION

MATERIALS COMPATIBILITY
   FUEL SYSTEM ELASTOMERS
   TURBINE BLADE ALLOYS & COATINGS

FUELS TOXICITY

ENGINE ENDURANCE   CS-77684

FORECAST OF FUTURE AVIATION FUELS

J. Morley English,

University of California, Los Angeles

FIVE SCENARIOS

1. OPTIMISTIC ECONOMIC GROWTH

2. MODERATE ECONOMIC GROWTH

3. INTERRUPTED ECONOMIC GROWTH

4. INSTITUTIONALLY CONSTRAINED GROWTH

5. RESOURCE LIMITS TO GROWTH

RANGE OF FUTURES

- INTERRUPTED GROWTH

    - ENERGY CONSTRAINED

- MODERATE GROWTH

    - ENERGY AVAILABILITY

- NO GAP

EFFECTS ON AVIATION

- CHANGED PETROLEUM SUPPLY

    1. COMPETING DEMANDS

    2. CHANGED SUPPLY

    3. GEOGRAPHICAL DISTRIBUTION

- SYNTHETICS

    1. OIL SHALE

    2. COAL

ISSUES

▲ PETROLEUM BASED FUEL

- GEOGRAPHICAL VARIATIONS

- MIXING

- TRANSPORTATION

- HYDROGENATION & HYDROCRACKING

▲ SYN FUELS

- COMBINATION WITH PETROLEUM

- RATE INTRODUCED

▲ ECONOMICS

- NEW INFRASTRUCTURE

- PRICE TRENDS

INTERRUPTED GROWTH

- DECISION TIME LAG

- CAPITAL IN INFRASTRUCTURE

- RESURGENCE OF GROWTH

- NEW EMERGENT LIFE STYLE

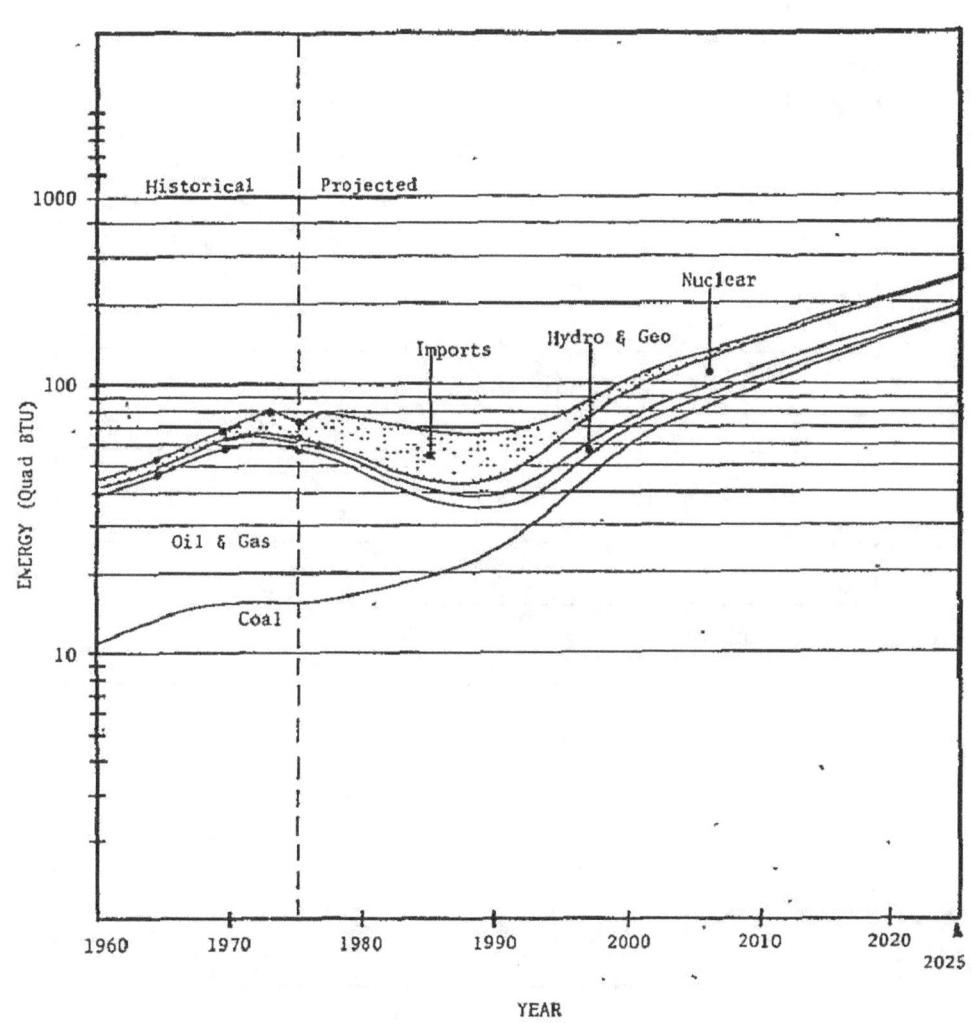

U.S. ENERGY SUPPLY -
HISTORICAL AND PROJECTED TO 2025
(INTERRUPTED SCENARIO)

UNINTERRUPTED GROWTH

- ENERGY WILL NOT CONSTRAIN

- GNP GROWS AT 3%

- U.S. ENERGY DEMANDS MET BY IMPORTS

- SYN FUELS WILL GROW RAPIDLY

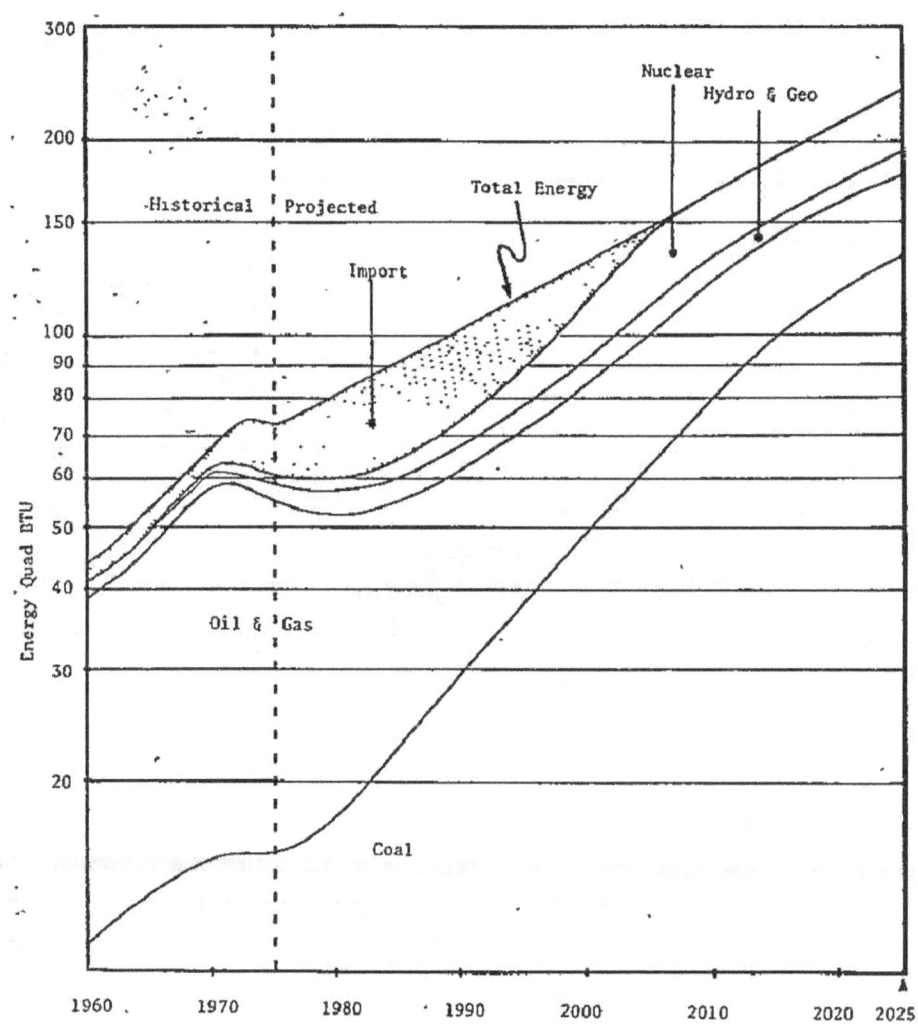

U.S. ENERGY SUPPLY-
HISTORICAL & PROJECTED TO 2025
(UNINTERRUPTED SCENARIO)

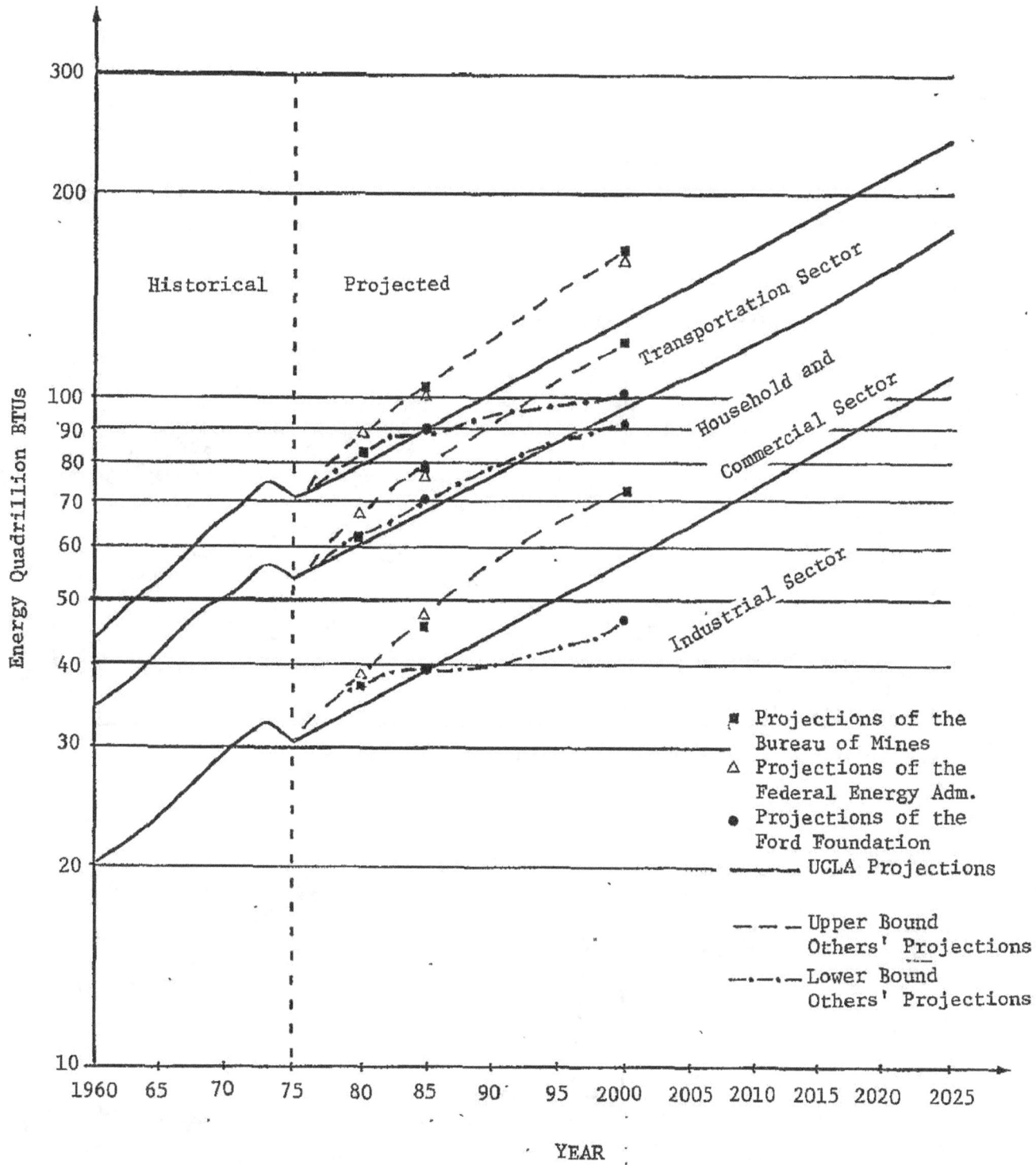

U.S. SECTORIAL GROSS ENERGY INPUT
(UNINTERRUPTED SCENARIO)

25

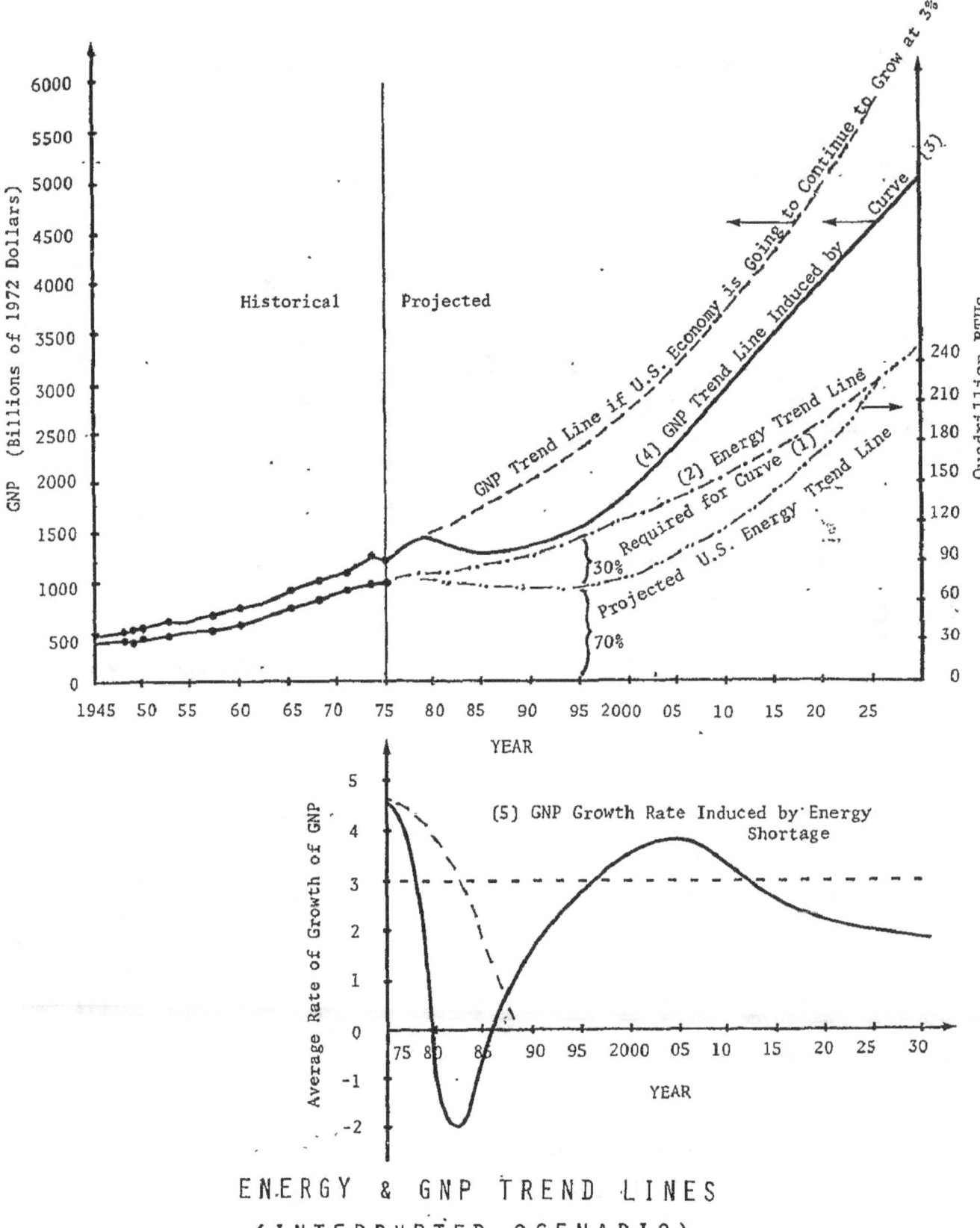

ENERGY & GNP TREND LINES
(INTERRUPTED SCENARIO)

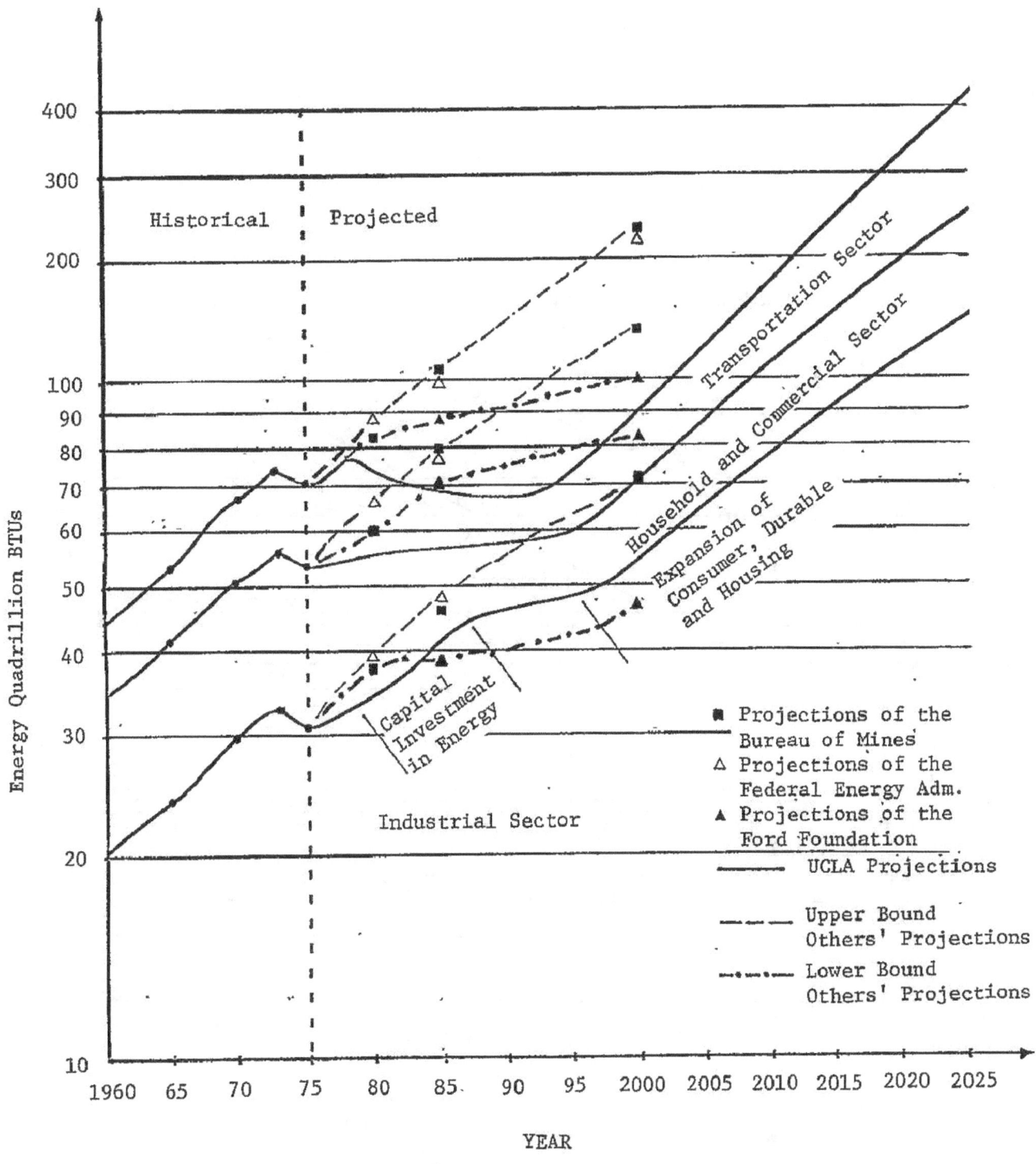

U.S. SECTORIAL GROSS ENERGY INPUT
(INTERRUPTED SCENARIO)

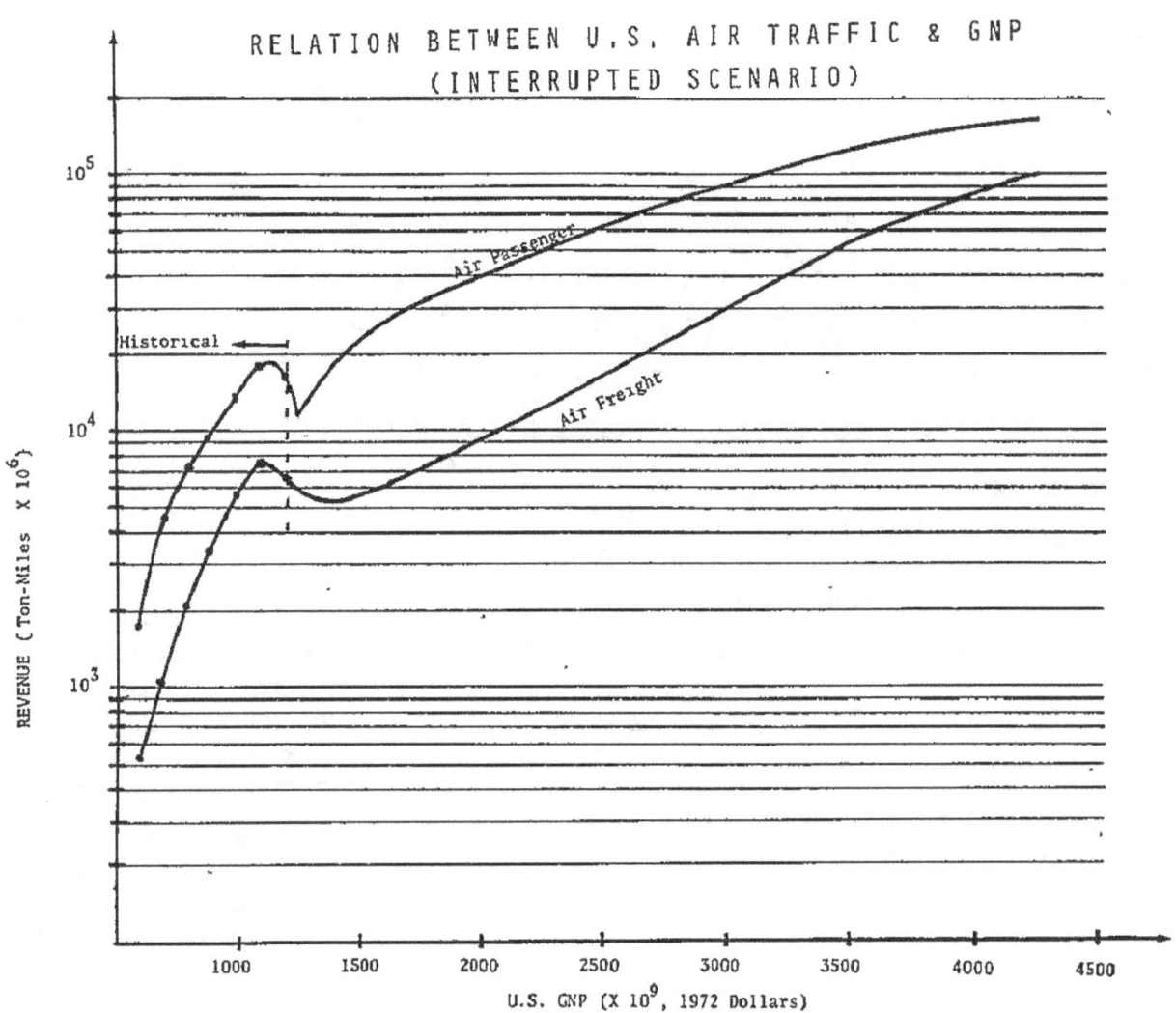

RELATION BETWEEN U.S. AIR TRAFFIC & GNP
(INTERRUPTED SCENARIO)

REVENUE (Ton-Miles X 10^6)

Air Passenger

Air Freight

Historical

10^5

10^4

10^3

1000 1500 2000 2500 3000 3500 4000 4500

U.S. GNP (X 10^9, 1972 Dollars)

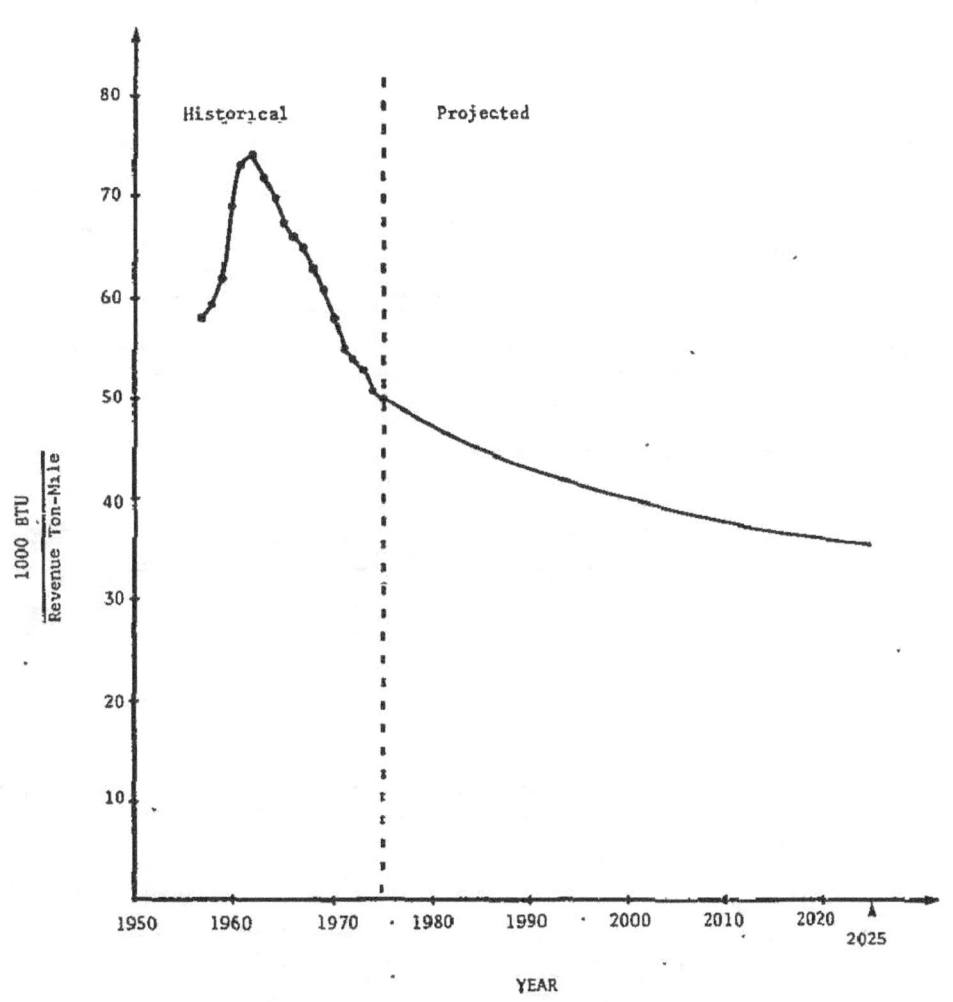

U.S. FLEET FUEL EFFICIENCY
CERTIFIED U.S. CARRIERS
(INTERRUPTED SCENARIO)

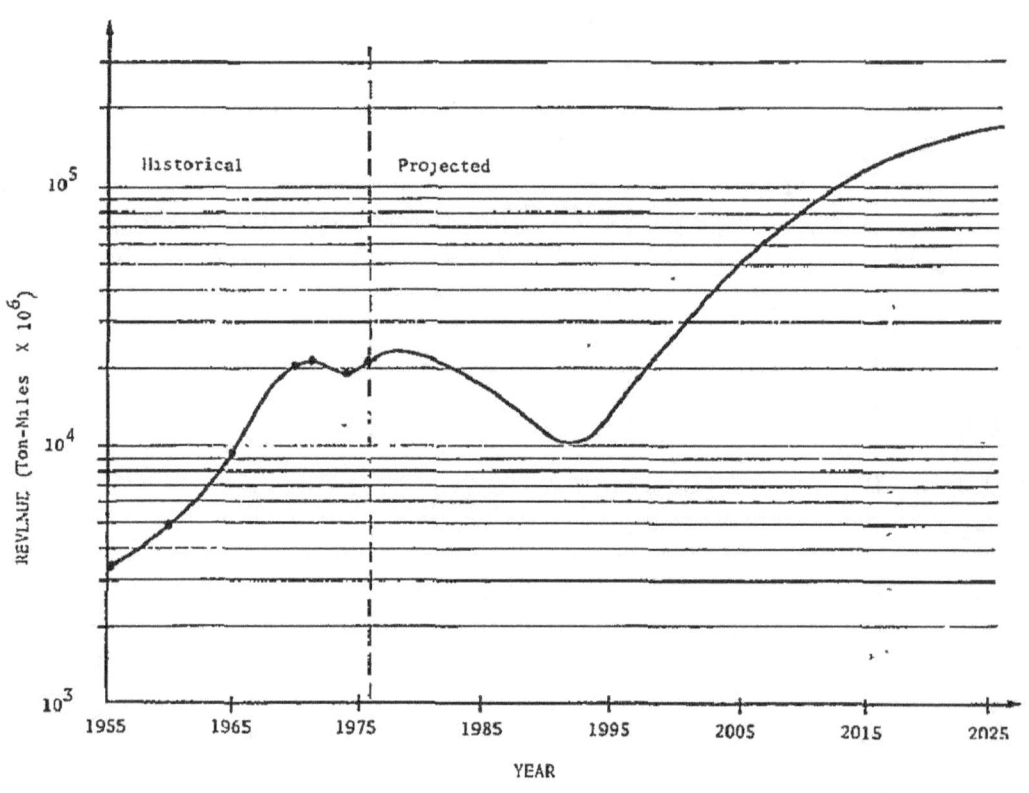

U.S. AIR PASSENGER REVENUE TON-MILES
(INTERRUPTED SCENARIO)

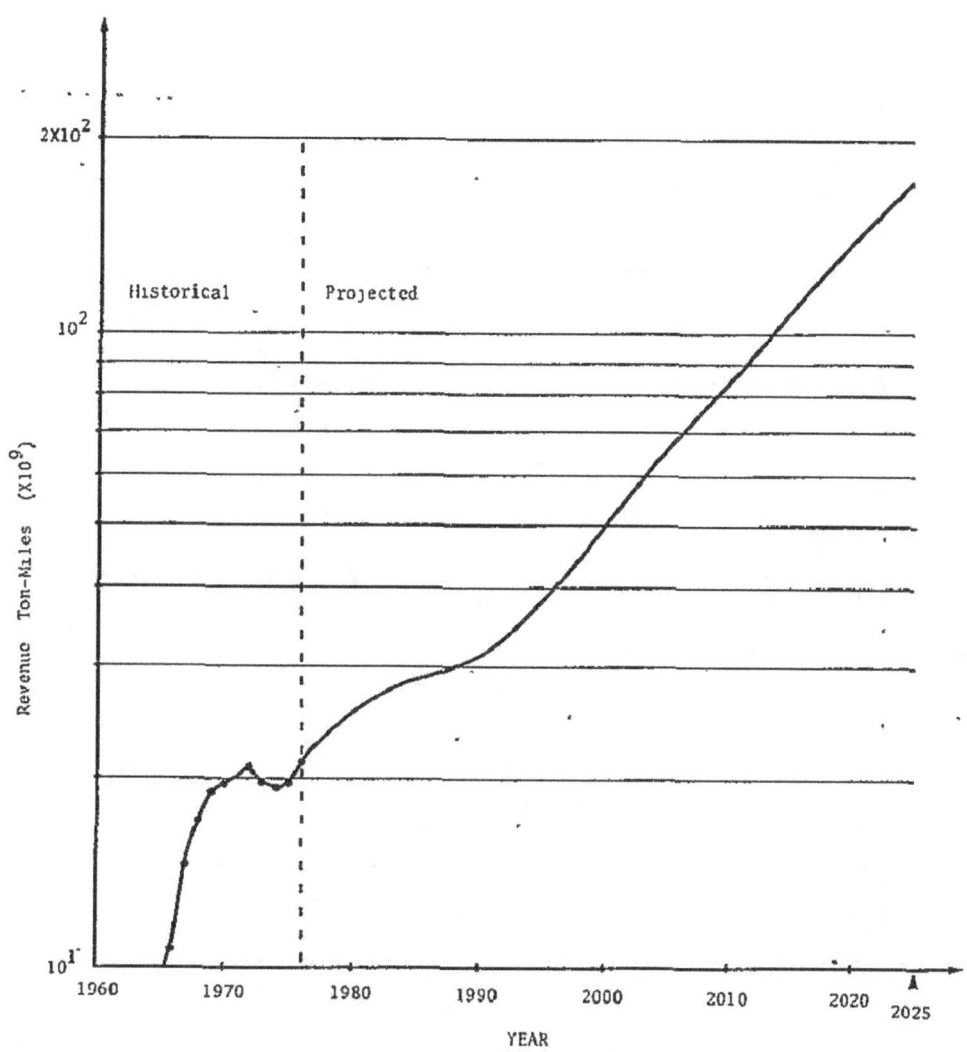

U.S. AIR PASSENGER REVENUE TON-MILES
(UNINTERRUPTED SCENARIO)

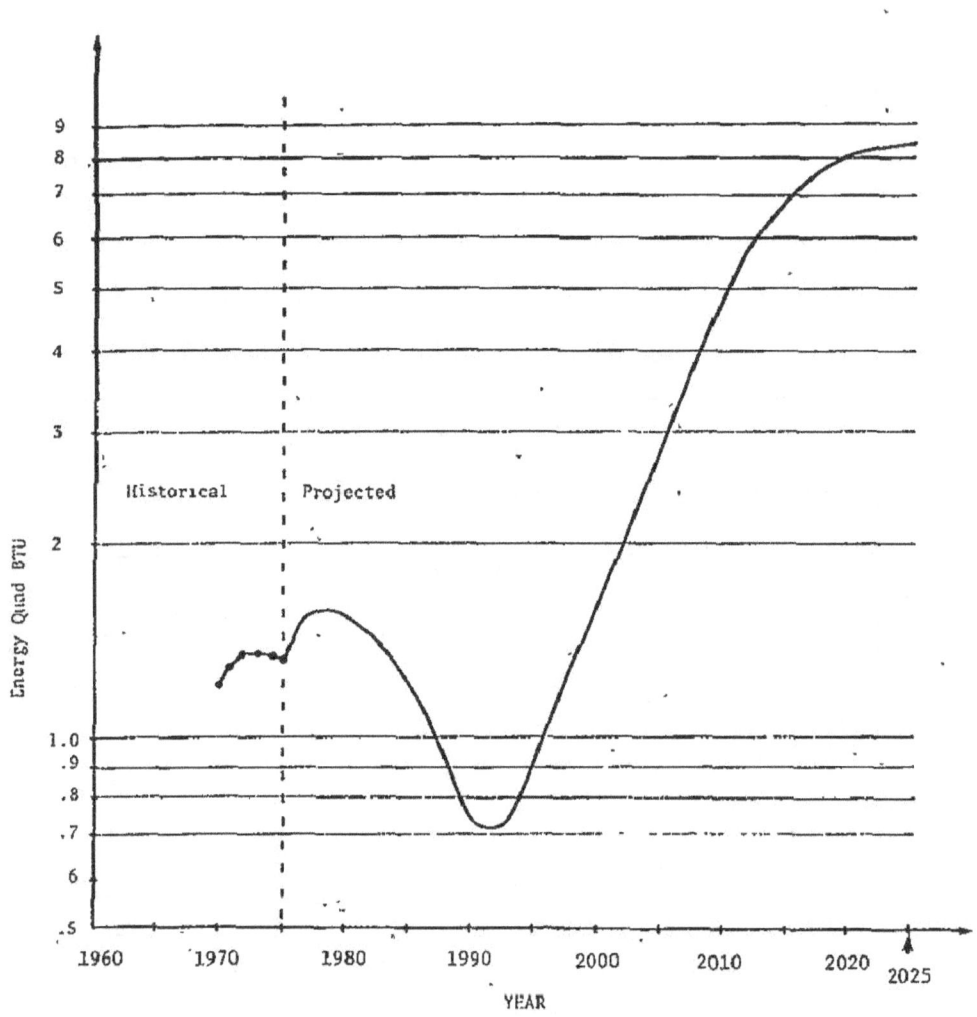

AIR TRANSPORTATION ENERGY REQUIREMENT
(INTERRUPTED SCENARIO)

TWO EXTREMES

HIGH FUEL   LOW MAINTENANCE

LOW FUEL   **HIGH MAINT**

GOAL ▪ MAXIMUM PROFIT

1976 MAINTENANCE COSTS	
TOTAL MAINTENANCE	200 $\overline{M}$
AIRCRAFT	103 $\overline{M}$
ENGINE	97 $\overline{M}$

1976 ENGINE PARTS COST	
ENGINE PARTS (INCL. REWORK LABOR)	97 $\overline{M}$
PARTS AFFECTED BY RADIATION	20.1 $\overline{M}$
WIDE BODY CF6, JT9D	15.3 $\overline{M}$
NARROW BODY JT8D, JT3D, JT4	4.8 $\overline{M}$

.33

FOR EACH GALLON OF FUEL CONSUMED,
HOT PARTS AFFECTED BY FLAME
RADIATION COST 1-1/3 CENTS IN 1976.
(20.1 M/1.5-B-GAL = 1.34 CENTS)

RATIO OF FUEL COST TO HOT PARTS COST
WAS APPROXIMATELY 25:1.

COST RATIO
FUEL/ENGINE MAINTENANCE

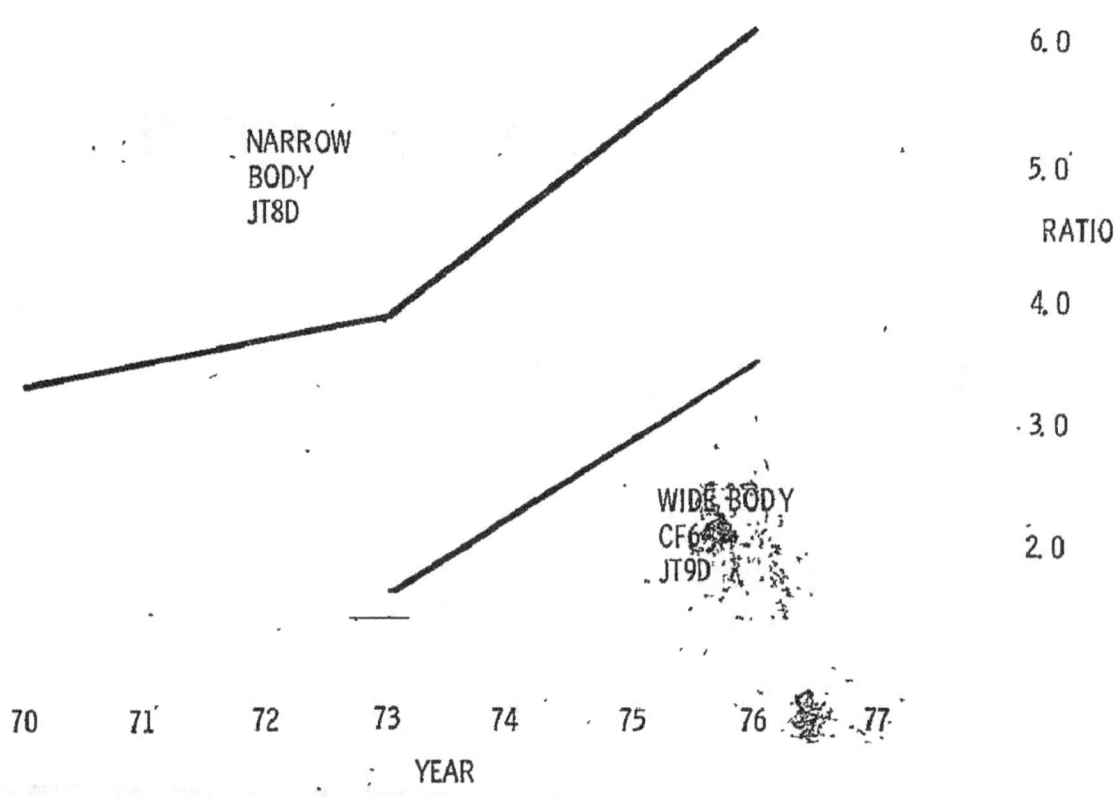

FUEL COST
1973 BASE

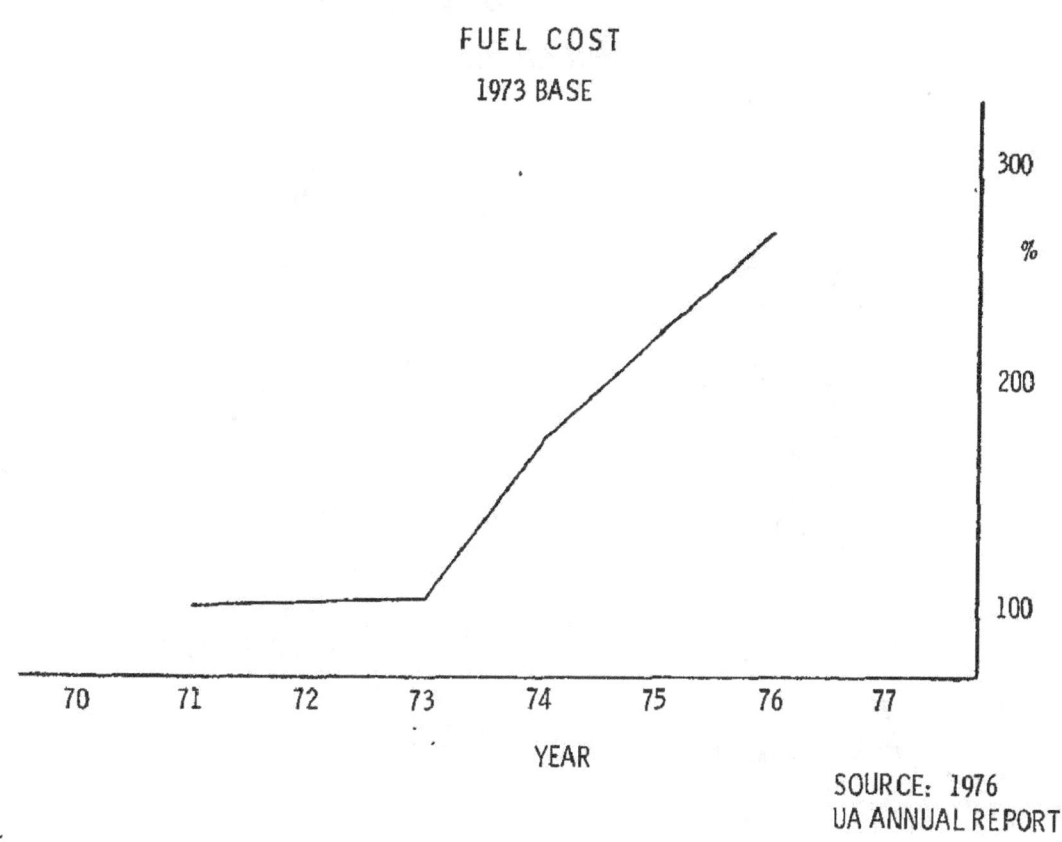

YEAR

SOURCE: 1976
UA ANNUAL REPORT

ENGINE MAINTENANCE COST

1974 BASE

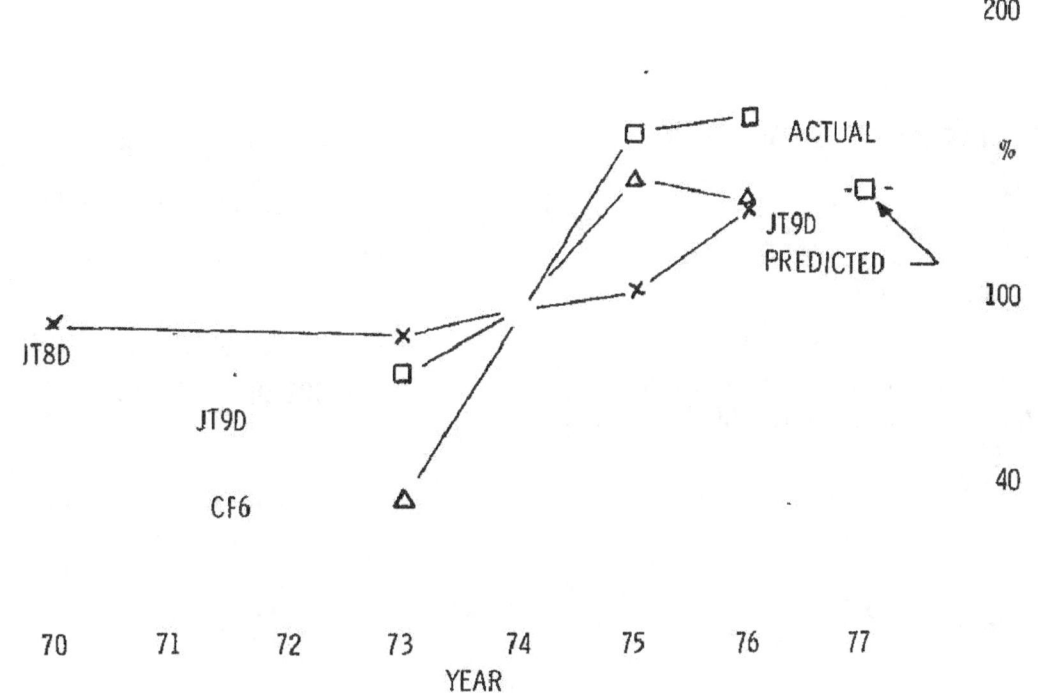

YEAR

TURBINE AIRCRAFT OPERATING COST

OUT-OF-POCKET

1976

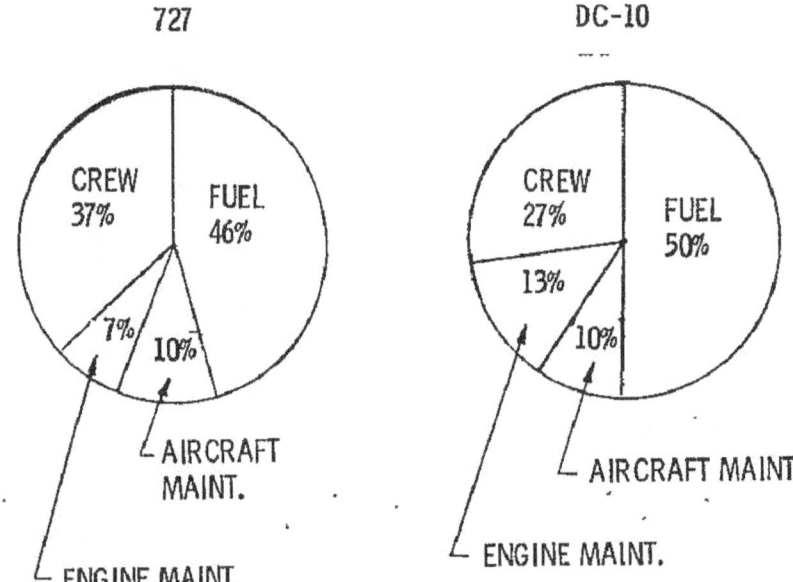

727   DC-10

EFFECT OF INCREASED AROMATICS

ON FUEL COSTS

1. POTENTIAL REDUCTION IN COST RESULTING FROM INCREASED AVAILABILITY.

2. POTENTIAL REDUCTION IN REFINING COST.

3. REDUCTION IN GALLONS REQUIRED DUE TO INCREASE IN BTU/GAL. (THIS COMPENSATES FOR LOWER BTU/LB.)

• EFFECT OF INCREASED AROMATICS ON AVAILABILITY VARIES BUT CAN BE QUANTIFIED.

• RESULTING EFFECT ON FUEL COST IS MORE DIFFICULT TO DETERMINE.

1976

FLEET	FUEL $	CONS. GALS.
WIDE BODY	144 $\overline{M}$	455 $\overline{M}$
NARROW BODY	336 $\overline{M}$	1.045 B
ALL	480 $\overline{M}$	1.5 B

DELTA HOT PARTS COST DUE TO INCREASED AROMATICS PROJECTED PROBABLY LESS THAN 5%. THIS IS EQUIVALENT TO 0.067 CENTS/GALLON OR $1 M/YEAR FOR UA.

A 1% INCREASE IN AROMATIC CONTENT IS CONSIDERED EQUIVALENT TO $0.5 M FUEL COST REDUCTION

SUMMARY

1. PRIMARY PROBLEM - ECONOMIC.
2. AIRWORTHINESS & SAFETY STANDARDS MUST BE MAINTAINED.
3. ABILITY TO TRACK IMPACT OF SPECIFICATION CHANGES REQUIRED.

ENGINE HOT SECTION PARTS AFFECTED BY RADIATION

PERCENT COST USED	JT9D	CF6	JT8D	JT3D
100%				
COMBUSTION CHAMBERS	X	X	X	X
1ST STG. TURB. BLADES	X	X	X	X
1ST STG. VANES	X	X	X	X
FUEL NOZZLES	X	X	X	
1ST STG. OUTER AIR SEAL			X	
COMBUSTION CH. OUTLET DUCTS AND CLAMPS			X	X
COMBUSTION CH. HEAT SHIELDS				X
50%				
2ND STG. TURB. BLADES	X	X	X	X
2ND STG. TURB. VANES	X	X	X	X

FUEL SUPPLIERS' PROBLEMS

William G. Dukek

Exxon Research and Engineering Company

FACTORS AFFECTING THE LONG

RANGE OUTLOOK

- TOTAL ENERGY DEMAND – WORLD AND U. S.
  DEPENDENCE ON PETROLEUM INTO THE 90'S
- SHIFT IN PRODUCT DISTRIBUTION IN U. S.
- CONVERSION PROCESSES FOR DISTILLATES
- OPTIMIZING AVIATION FUEL IN THE 90'S

WORLD ENERGY DEMAND

(EXCLUDING COMMUNIST AREAS)

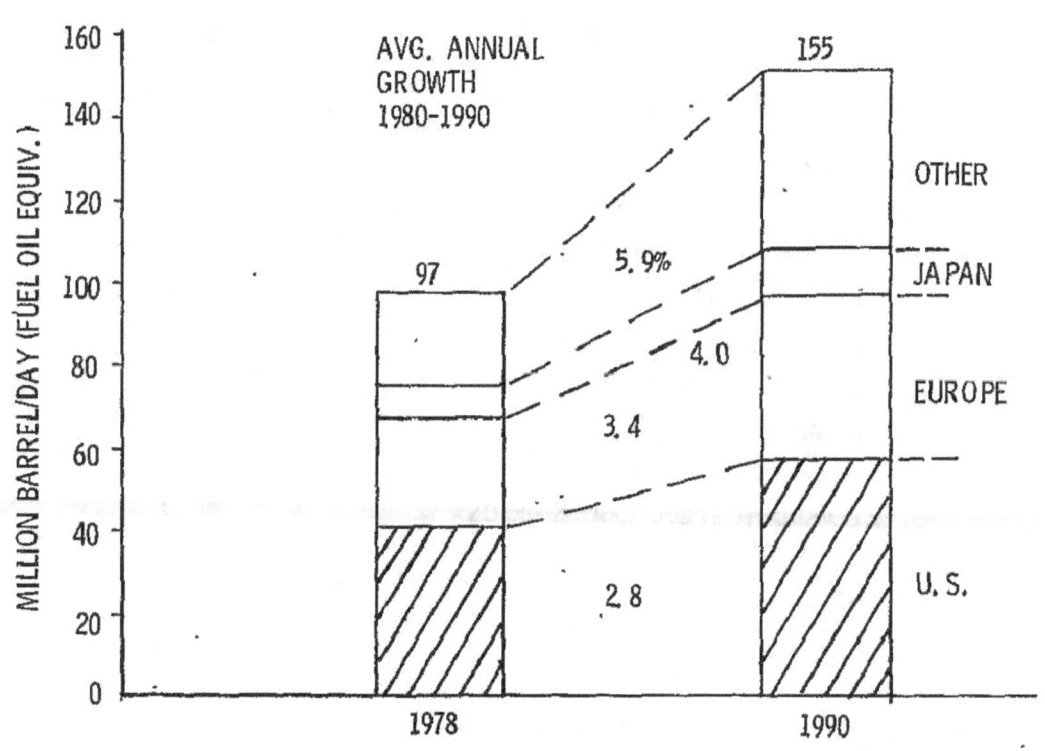

HOW 1978-1990 ENERGY GROWTH WILL BE MET

(MILLION BARREL/DAY (FUEL OIL EQUIV.))

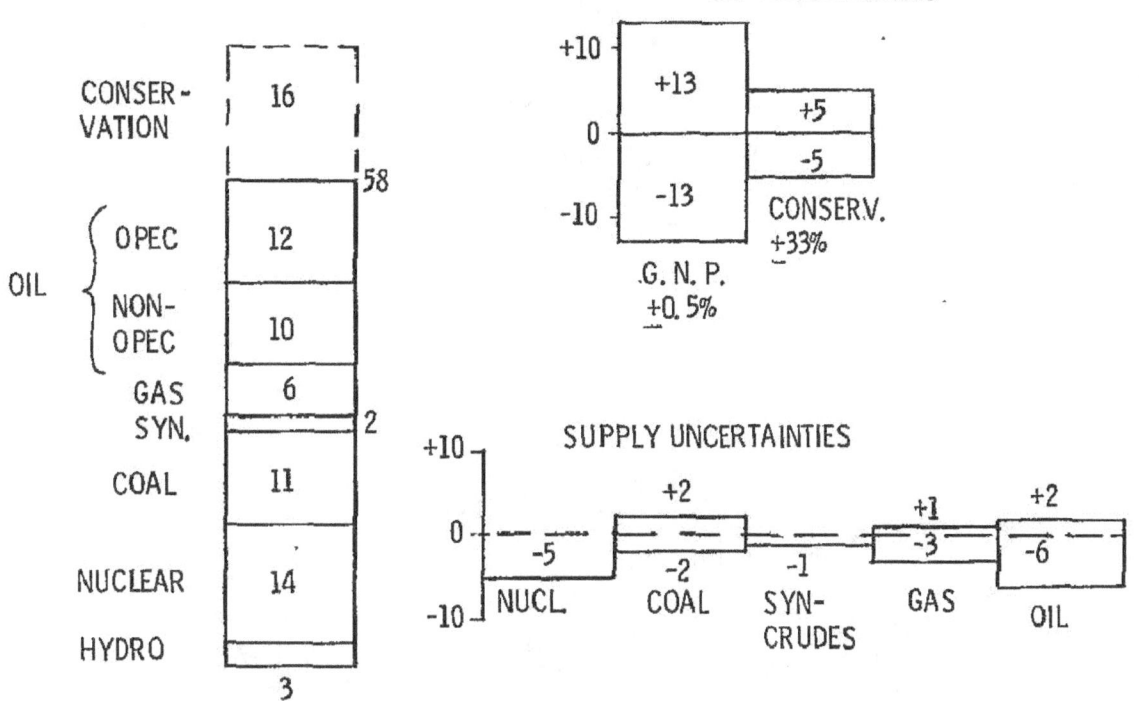

DEMAND UNCERTAINTIES

SUPPLY UNCERTAINTIES

WORLD OIL SUPPLY

(EXCLUDING COMMUNIST AREAS)

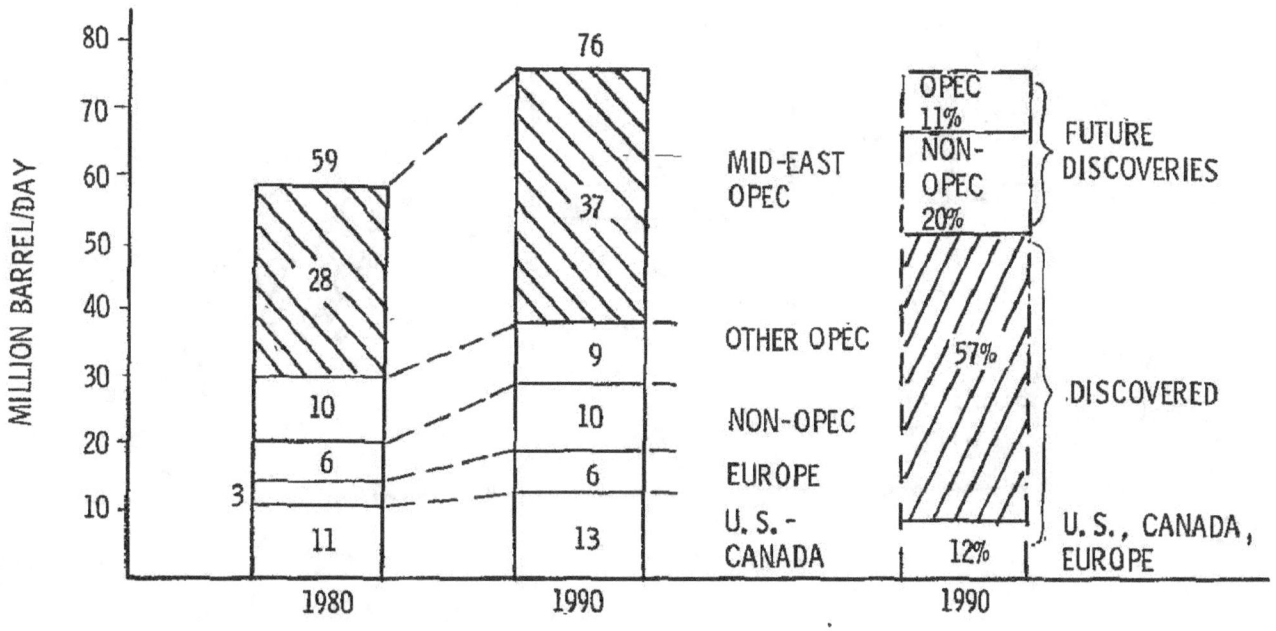

PROBLEMS FACING FUEL SUPPLIERS

IN THE 90'S

1. UNCERTAIN CRUDE SUPPLIES -
   IMPORTS STILL ~ 50%
   QUALITY OF MARGINAL CRUDES

2. REFINING CAPACITY

3. SHIFTS IN PRODUCT DEMANDS

4. COMPETITION FOR DISTILLATES

PRODUCT YIELD VS. DEMAND

(BASIS: 100 000 BARREL/DAY U. S. REFINERY)

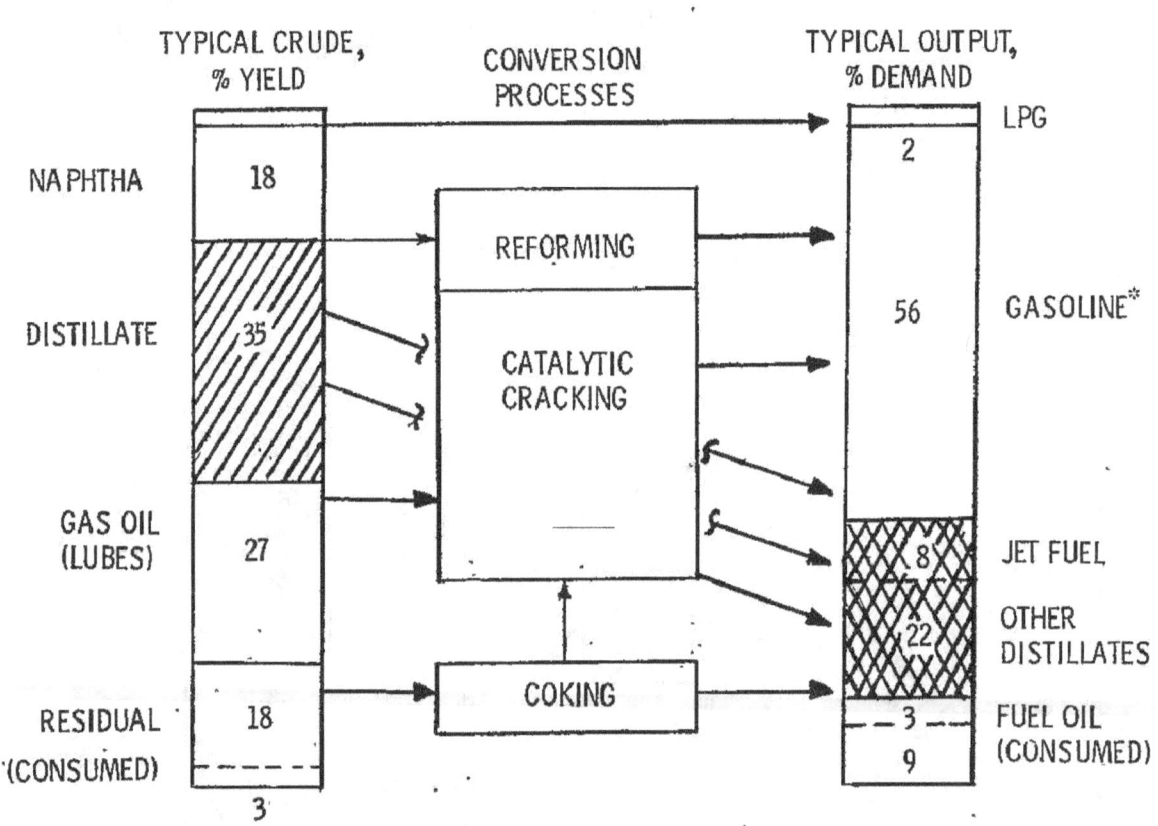

*GAS/DIST. RATIO = 1.7

U. S. PETROLEUM PRODUCTS DISTRIBUTION

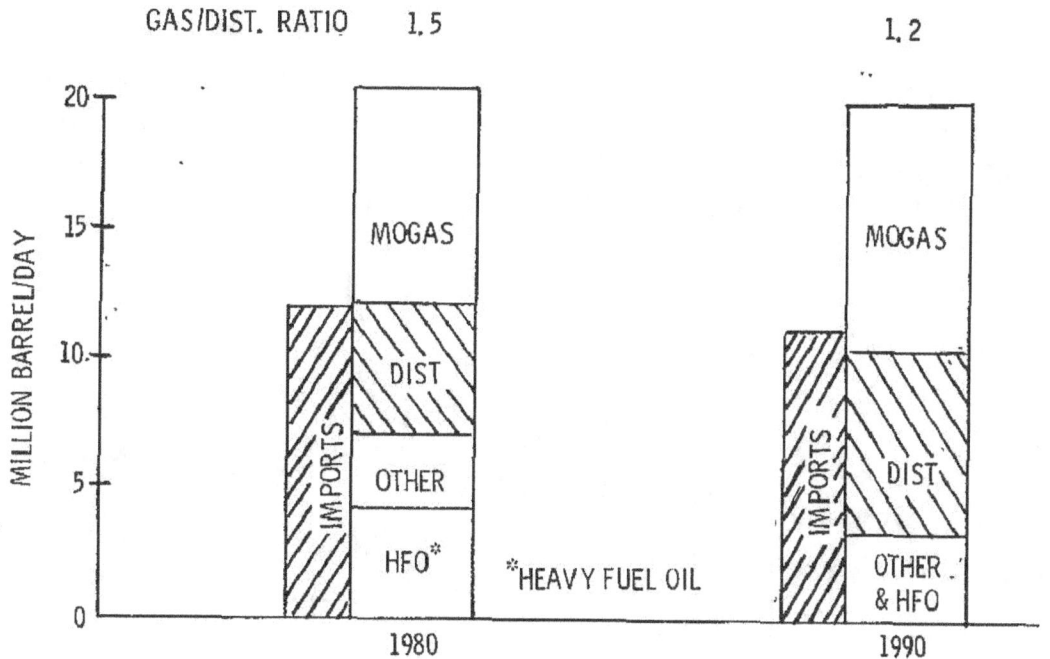

MAXIMIZING DISTILLATES IN FUTURE REFINERY

(BASIS: MOGAS/DIST.RATIO = 0.7)

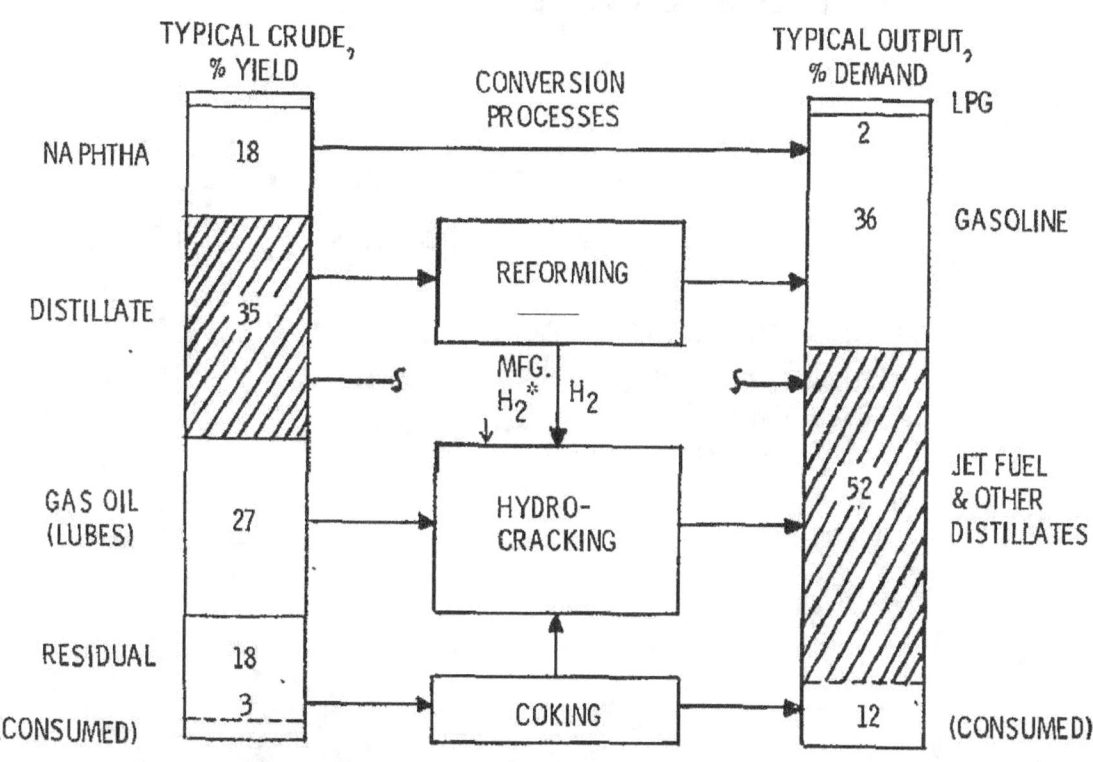

*ADDITIONAL H_2 TO BE SUPPLIED FOR HYDROCRACKING IN ADDITION TO REFORMER BYPRODUCT

BLENDING DISTILLATES IN FUTURE REFINERY

BLENDING STOCKS
(% YIELDS SHOWN)

DISTILLATE PRODUCTS

EXPERIMENTAL
REFEREE BROAD-
SPECIFICATION
FUEL

DISTILLATION
CUTS

HYDRO-
CRACKATE

JET A

°C.

DISTILLATION RANGE

KEROSENE
(20% AROM.)

9.8%

LT. VAC. GAS OIL
(30% AROM.)

12.5%

29.7%

H/T*

H₂

— 150

— 200

— 250

— 300

— 350

*AROMATICS REDUCTION TO MEET JET A SPECIFICATION BY HYDROTREATING

SUPPLEMENTING KEROSENE FOR FUTURE JET FUEL

CUT	AROMATICS, %	BOILING RANGE, °C
PIPE-STILL DISTILLATE SIDE STREAM	15-30	230-350
HYDROTREATED CATALYTIC CRACKED DISTILLATE (OLEFIN SATURATED)	30-40	200-350
HYDROCRACKED DISTILLATE	10-30	200-350
HEAVY VIRGIN NAPHTHA	10-20	130-200

SUPPLEMENTARY JET FUELS IN TERMS OF ENERGY COST

PRIORITY	CUT	CRITICAL PROPERTY	% ENERGY REQUIRED
1	PIPE-STILL DISTILLATE	AROMATICS, FREEZE POINT	3
2	HEAVY VIRGIN NAPHTHA	FLASHPOINT	3
3	H/T CAT DISTILLATE	AROMATICS	6
4	HYDROCRACKATE	AROMATICS (?)	12
5	H/T HYDROCRACKATE	------	20

SUMMARY OF FUEL SUPPLIERS OUTLOOK

- DEPENDENT ON IMPORTED AND MARGINAL CRUDES
- SYNTHETIC CRUDES WILL BE MINOR COMPONENT
- U. S. REFINERS MUST MAXIMIZE DISTILLATES
- CONVERSION PROCESSES ARE ENERGY INTENSIVE
- SPECIFICATION JET FUELS REQUIRE AROMATICS REDUCTION
- PROPOSED BROAD SPECIFICATION FUEL CAN REPRESENT OPTIMUM DISTILLATE FOR REFINERY FLEXIBILITY AND MINIMUM COST

APPENDIX B

WORKING GROUP REPORTS

WORKING GROUP I - AVIATION FUELS SUPPLY AND DEMAND

General Findings and Conclusions

1. The demand and supply outlook for aviation fuels in the 1990's reinforces the proposals that new aircraft should be capable of operating on fuel that is wider cut than Jet A kerosene. A broad-specification, petroleum-based fuel similar to No. 2 diesel provides refiners with the maximum flexibility for meeting jet fuel demand with the least costly processing sequence.

2. Such a fuel can be introduced into the world and U.S. market in several ways. Finding the optimum way requires study of current aircraft and ground distribution systems and trade-offs between fuel manufacturing problems and airline penalties.

Specific Recommendations for Studies

1. Current aircraft. - To what extent and at what cost can remaining current-model aircraft be retrofitted, beginning about 1990, to use broad-specification diesel-type fuels?

2. Ground distribution systems. - What is the investment/operating cost comparison between a system-wide introduction of a new jet fuel in the United States in 1990 and the introduction of a separate distribution system for only domestic use of the broad-specification fuel at that time (i.e., maintaining Jet A for international and other special operations)?

3. Cost/penalty trade-offs. - How is the relative cost advantage of a broad-specification fuel (compared with Jet A fuel) best assessed against the penalty of operating new aircraft or retrofitted aircraft adapted for such a fuel?

Additional Observations and Study Recommendations

1. Trade-off studies can be made against a standard specification or reference specification, but the panel was not prepared to recommend a particular broad-specification fuel as the standard.

2. While it is impossible to know the characteristics of future oil supplies, crudes supplied from presently known reserves may be expected to yield an increasing propor-

tion of aromatics. It will be prudent to assume that this trend to higher aromatic fuel will continue for fuels coming from newly discovered fields. Therefore, the aircraft operator will be faced with the alternatives of higher maintenance costs or higher prices for fuels. (The cost of some form of hydrogen treatment for limiting aromatics will increase fuel cost.) Economic trade-off studies should be undertaken to determine the most economic aromatic level for various fuel scenarios of the 1990's. Coupled with this problem of aromatics will be the problem of thermal stability.

3. Engineering economic studies should be made to determine the optimum freezing point for the new broader specification fuels of the 1990's in terms of trade-offs between heating system costs and improved price/supply relationships. These studies should include considerations of conditions under which operations may be modified (e.g., special supply and summer/winter variation in operating policy). Such modified practices may be used to cope with the problem.

4. It is undesirable and perhaps impossible to provide dual fuel supplies at airports. However, fuels of different specifications may be supplied in different geographical locations and, therefore, the range of effects of geographical variations in fuels should be evaluated.

5. The target fuel specification for aircraft of the 1990's should be determined as a basis for new aircraft design, but the transition to such fuels should be incremental in order to minimize the impact of retrofit costs.

6. Economic trade-off studies should be performed because of the competition for fuel supplies by other users, particularly from other transportation modes. A significant part of fuel cost is that of supply and distribution. Therefore, serious consideration should be given to a common standardized fuel that, while offering promise for expanded supply at relatively lower cost, would need to have a broader range of properties.

7. The scenario approach should be extended to include supply and demand situations incorporating hypothetical crudes and hypothetical geographical locations to evaluate economical supply mixes for the demand side.

8. A study should be undertaken in conjunction with the UCLA scenario models to determine the timing for introducing changes. Based on this study, a recommendation should be made for effecting the required changes.

WORKING GROUP II – COMBUSTOR RESEARCH AND TECHNOLOGY

FOR BROAD-SPECIFICATION FUELS

General Findings and Recommendations Concerning Use of

Broad-Specification Fuels

Findings:

1. With any given level of combustion design technology, higher levels of carbon monoxide, hydrocarbons, oxides of nitrogen, and smoke emissions will be obtained with broad-specification fuels than with present-day jet fuels. Appropriate adjustments of the EPA emissions standards to accommodate these new fuels will probably be needed. The EPA should be continually provided with pertinent data so that appropriate and timely actions can be taken to define needed adjustments to the standards.

2. The prescribed nitrogen content limits (0.005 percent) of a tentative fuel specification are quite low and, therefore, preclude any NO_x emissions concerns from this source. From a combustor design standpoint, this limitation is advantageous, since no promising combustor design concepts for suppressing NO_x emissions due to fuel-bound nitrogen have been identified.

3. With any fuel, viscosity must be limited to a maximum value of about 15 centistokes in order to obtain acceptable ignition and altitude relight performance. With the broad-specification fuels, viscosity characteristics are at or near the maximum allowable values, and this 15-centistoke viscosity requirement will limit the minimum fuel temperatures at the combustor inlet to a range of about 10^0 to 40^0 F. On cold days, fuel heaters will be needed at starting conditions with such fuels. Also special provisions to permit satisfactory ignition at altitude relight operating conditions will be needed.

4. Currently used combustor component testing methods (subassembly tests, sector combustor tests, and full annular combustor tests) are expected to be adequate for the development of combustors intended for operation with the new fuels.

5. To permit the development of improved combustor design models and analysis techniques and improved correlations of fuel properties with combustion characteristics, additional data on the basic combustion and atomization characteristics of low-hydrogen-content fuels are needed. Thus, additional laboratory investigations, along the lines of the NASA soot formation/oxidation program and the NASA Stratospheric Cruise Emissions Reduction Program (SCERP) investigations of autoignition phenomena, should be considered to provide the needed basic data.

6. The possibility of defining a highly standardized and precisely characterized fuel blend for use in developing combustors intended for low-hydrogen-content fuels should be considered. Preferably, this blend should consist of known amounts of prescribed hydro-

carbon constituents - such that this standardized blend could be obtained by all groups involved in alternative fuels development efforts. If feasible, the use of such a referee fuel (or calibration fuel) would be expected to result in more uniform and representative results among all investigators. The proposed fuel specifications appear somewhat too broad in some important properties. The relatively wide allowable variations in these properties might prevent the attainment of consistent data on the effects of using the new fuels.

7. More accurate methods of measuring the key fuel constituents are critically needed in order to characterize fuels. In particular, more precise methods of measuring fuel hydrogen contents are needed since hydrogen content is believed to be the most significant fuel characteristic in terms of defining its combustion properties. Capabilities for measuring fuel hydrogen content within ±0.02 percent are needed.

8. As an ongoing effort, programs to develop fire-safe fuels should be monitored. As appropriate, the results of these programs should be factored into the overall alternative hydrocarbon fuels development programs.

Recommendations for additionally needed technology development efforts:

A-1. Initiate programs to provide more complete and improved fundamental data on the combustion and atomization characteristics of low-hydrogen-content fuels.

A-2. Define and develop improved methods of measuring the properties and chemical constituents of fuels, especially hydrogen.

Specific Findings and Recommendations Concerning Use of Broad-Specification

Fuels in Current-Technology Combustors

Findings on impacts:

As compared with current jet fuels, the use of the proposed fuels is expected to result in

1. Significantly increased smoke levels because of the decreased fuel hydrogen contents.

2. Significantly decreased altitude ignition capabilities because of the increased fuel viscosity and decreased volatility characteristics

3. Significantly increased metal temperatures because of the decreased fuel hydrogen contents

4. Increased carbon deposition tendencies because of the decreased fuel hydrogen contents

5. Modestly increased carbon monoxide and hydrocarbon levels at idle because of the increased fuel viscosity and decreased volatility characteristics

6. Slightly increased NO_x levels at high power because of the decreased fuel hydrogen contents

7. Probably increased fuel nozzle plugging and gumming problems because of the decreased fuel thermal stability properties.

Findings on pertinent design and development actions:

1. The specific impacts of using the proposed fuels are expected to be significantly further quantified as a result of ongoing NASA combustor design study and combustor rig test programs and by the similar USAF-sponsored fuel effects programs that involve combustor rig tests.

2. Some combustor design modifications will probably be needed to accommodate the new fuels satisfactorily.

3. The use of increased combustor dome airflow quantities to reduce the expected higher smoke levels with broad-specification fuels will probably further reduce altitude ignition capabilities. The alternative approach of using decreased dome airflows to increase altitude ignition capabilities will very probably further increase smoke levels.

4. The use of increased liner cooling flows to reduce the expected higher liner metal temperatures with the proposed fuel will probably result in some degradation in exit temperature profile and pattern factors.

Recommendations for additionally needed technology development efforts (in order of priority):

B-1. To further extend the data base being obtained in the various ongoing NASA, USAF, and USN programs, conduct rig tests of additional combustors, including business jet engine combustors, with various alternative hydrocarbon fuel blends.

B-2. Verify and extend at least some of these resulting combustor rig test data in engine tests.

B-3. Based on these combustor and engine test results, define any needed adjustments to the proposed fuel specifications.

B-4. Initiate programs to define and develop technology for obtaining acceptable ignition and altitude relight capabilities, with the new fuels, in combustors designed to operate with acceptable smoke levels. Specific approaches to be investigated include alternative ignition methods, added fuel control system capabilities, alternative fuel injection methods, and fuel staging methods.

B-5. Initiate programs to define and develop technology for maintaining acceptable dome and liner metal temperatures, with the new fuels, without reliance on large increases in cooling airflows. Specific approaches to be investigated include thermal barrier coatings and added cooling provisions.

Specific Findings and Recommendations Concerning Use of Broad-Specification

Fuels in Advanced-Technology Combustors

Findings on impacts:

1. With NASA Experimental Clean Combustor Program (ECCP)- type staged combustors, the impacts of using broad-specification fuels are expected to be similar to those in current-technology combustors, with the following exceptions:

a. The lean main-stage operation of staged combustors at high power may minimize the smoke level increases and the associated higher metal temperature problems expected with the use of these new fuels in current-technology combustors. Currently available test results, however, are inconclusive. Similar comments are expected to apply to variable-geometry combustors designed to operate with reduced pollutant emission levels.

b. The generally more favorable combustion zone stoichiometries and flow conditions obtained in staged combustors at lightoff and low-power conditions, due to the use of pilot-stage-only operation at these conditions, may minimize the ignition and altitude relight problems expected with the use of these new fuels in current-technology combustors. Again, currently available test results are inconclusive. Similar comments apply in the case of variable-geometry combustors designed to operate with reduced pollutant emission levels.

c. Fuel nozzle plugging problems will probably be more severe in staged combustors because of the significantly greater number of fuel injectors used in these advanced combustors and the need to shut off some injectors at some engine operating modes.

2. In the case of the very advanced combustor design concepts (premixing/ prevaporizing combustors, with and without catalytic reactors - as in SCERP), the various adverse impacts of the use of the proposed fuels are expected to be less significant than those in conventional and staged combustors. However, other fuel characteristics, such as flashback and preignition, may compromise the use of the broad-specification

fuel in these advanced combustor design concepts. In any event, the feasibility of using these very advanced combustors in engine applications has not yet been established, even with current fuels.

Findings on pertinent design and development actions:

1. NASA-sponsored combustor design studies are expected to somewhat quantify the specific problems and impacts associated with the use of these new fuels in NASA ECCP-type staged combustor designs.

2. SCERP autoignition/flashback investigations will provide key data on the problems associated with the use of alternative fuels in premixing combustors.

Recommendations for additionally needed technology development efforts (in order of priority):

C-1. Conduct additional tests of existing staged combustors with low-hydrogen-content fuels.

C-2. As required, develop the needed design improvements to accommodate the use of the proposed fuels in these existing staged combustor design concepts.

C-3. Include the use of the proposed fuels as an integral part of all future staged, variable-geometry, and/or SCERP combustor development programs.

Summary of Key Recommendations for Needed Near-Term

Technology Development Efforts

1. Conduct additional tests of existing NASA ECCP-type staged combustors with low-hydrogen-content fuels. As required, develop the needed design improvements to accommodate the use of these new fuels in these staged combustor design concepts.

2. Include the use of broad-specification fuels as an integral part of all future staged, variable-geometry, and/or SCERP combustor development programs.

3. Conduct additional and more extensive rig tests of existing current-technology combustors, including business jet engine combustors, with low-hydrogen-content fuels. Verify and extend at least some of the results of these rig tests in engine tests.

4. Define and develop improved methods of measuring the properties and chemical consituents of fuels, especially hydrogen.

WORKING GROUP III - FUEL SYSTEM RESEARCH AND TECHNOLOGY

FOR BROAD-SPECIFICATION FUELS

Goals and Approach

Working Group III members were requested to identify fuel system research and technology efforts needed to support the introduction of broad-specification fuels. The discussions were limited to the aircraft fuel system, the ground handling system, and the engine fuel system where it might not be reviewed by Working Groups II and IV.

The panel identified the following goals of the discussion:

1. Recommend research and development for approaching the problems of broad-specification fuels with the ground handling and airframe fuel systems.

2. Define a referee broad-specification fuel(s) for use in any test program.

The approach used by the panel was to identify property changes that occur with broadening the Jet A specification: lowering the initial boiling point in one case, and raising the final point in the second case. From this list a second "short list" of properties was developed that were considered to be worthy of more discussion. This list is shown at the end of this discussion.

In the case of lowering the initial boiling point it was agreed that any shift from Jet A would be small and definitely would not result in a fuel that had a Reid Vapor Pressure (RVP) greater than Jet B. Based on this opinion, it was agreed that there were no significant problems in lowering the initial boiling point, other than the relationship of vapor pressure to safety, which was the main topic of discussion of Working Group V and was therefore not addressed by this working group.

In the case of raising the final boiling point there were several property changes that were deemed to be significant.

Recommended Research and Development

Freezing point. - In general, it was considered that NASA's program in this area was satisfactory. Particular recommendations were as follows:

1. A follow-on study to the recently completed Boeing program (NASA CR-135198) should be made to refine and detail heating systems that will permit the use of high-freezing-point fuels. The follow-on study should include evaluation of the number of flights versus the minimum freezing point and the cost of operation versus the freezing point.

2. Some effort should be expended to support the business jet sector of the industry, who have little or no say in the fuel properties but are developing very long-range business jets.

3. A study should be made of fuel freezing behavior in a fuel tank (currently underway as NASA Contract NAS3-20814). The study may determine if two-phase flow can be tolerated. Results should be aimed at a refinery/laboratory test for flow ability and pumpability.

4. Full-scale testing should be done on a ground fuel system simulator, followed by a flight test (particularly if two-phase flow is shown to be feasible in the small-scale test).

5. A user "quick test," taking less than 10 minutes, would be useful to the operators. It could help them avoid diversions or optimize flight plans against the actual freezing point rather than specification values. Some ideas for quick tests were mirror-dewpoint apparatus with a thermal electric cooler, a cooled bar inserted in the fuel to measure wax buildup, and analysis of normal paraffin content by molecular sieve absorption.

6. An analysis of in-flight temperatures compiled by IATA is desirable.

Aromatics. - The compatibility of elastomeric materials as a function of aromatic content, up to 30 or 40 percent, should be investigated. A correlation to hydrogen content or some other property would also be desirable in view of the range of aromatic compounds that might be presented in a high-freezing-point fuel. One panel member suggested that any compatibility testing be conducted with a specific aromatic content defined for each portion of the distillation range of the test fuel. The work should include documentation of current materials used in commercial aircraft.

Water solubility and cleanliness. - The trends of these problems must be looked at, particularly with respect to aromatic content.

Viscosity. - This is not a problem in the fuel system unless the flow becomes non-Newtonian at very low temperatures, which will be examined as part of the freezing-point program. However, it was noted that viscosity must be examined carefully with respect to engine starting.

Other properties. - Except for those listed above, it was considered that the effects of fuel properties could be evaluated by available methods and that NASA research programs were not needed.

Referee Fuel

There was general agreement that the trend of fuel specification limits in the next 25 years would be to hold the initial boiling point near its current limit but let the final boiling point increase. On this basis, it was considered that it would be extremely unlikely that a 0° F freezing point fuel would be a reality. This was emphasized by the large operating cost increase noted in the Boeing study for a 0° F as compared with a

-20° F freezing point fuel. The panel therefore considered that a reference fuel with a freezing point near -20° F would be most representative of future fuels (through the year 2000) but that some work with a 0° F freezing point fuel would be useful to permit interpolation of data between -20° F and 0° F.

FUEL PROPERTIES CONSIDERED

LOWER IBP	HIGHER FBP
DENSITY	VISCOSITY
VAPOR PRESSURE	DENSITY
BTU/GALLON	FREEZING POINT
LUBRICITY	BTU/POUND
GAS SOLUBILITY	AROMATICS
	CLEANINESS
	WATER SOLUBILITY

WORKING GROUP IV - FUEL THERMAL STABILITY

RESEARCH AND TECHNOLOGY

General Findings and Conclusions

The characteristic of fuel known as thermal oxidation stability manifests itself in subsonic aircraft engines by deposits formed in the fuel nozzles or in the fuel lines between the fuel controller and the nozzles. These deposits can cause distortion of the nozzle spray pattern, which in turn can cause turbine damage. Deposits in other parts of the aircraft fuel system, such as fuel-oil heat exchangers and fuel controllers, do not limit operation with current engines and fuels.

Thermal stability is of significant concern with today's fuels and engines. The magnitude of the problem is due to a complex combination of variables such as engine model, length of flights, fuel contamination, and the average stability of the fuel used. Because of the critical nature of the current thermal stability problem and the anticipated higher fuel temperatures in nozzles of future engines, lower thermal stability specification limits would not appear practical.

Fuel thermal stability is so dependent on trace materials present in the fuel that no meaningful specification can be written for a referee fuel. Based on limited data, thermal stability of No: 2 diesel fuels will range from about 400° F to less than 500° F breakpoint temperature by the JFTOT test procedure. Recirculation of hot fuel back to the tank may reduce deposits caused by fuel instability. The possible benefits will depend on the specific design and whether or not recirculation is continuous.

Specific Recommendations for Study

The following areas are recommended for additional research:

1. There is a real need for a relationship between laboratory thermal stability test results and full-scale engine deposit results. Therefore, a research program that uses a fuel system simulator capable of covering all flight conditions is suggested in order to identify the sensitivity of the system to deposit formation. Results would aid in designing for minimum deposit formation and in predicting the effect of fuel quality. In essence, the simulator should be sized to be representative of a full-scale engine. To simplify the design, the requirements for takeoff capability might be compromised relative to air and fuel flow rates because nozzle fouling is minimal under takeoff conditions.

2. There should be continuing basic research on the controlling processes in deposit formation, such as surface material and fluid mechanics.

3. Fundamental laboratory studies should be extended to include the investigation of the chemical composition of the fuel so as to identify the chemical species deleterious to thermal stability. In particular, those materials should be identified that would occur from using fuel fractions heavier than those in current kerosene-type jet fuels.

4. Deoxygenation and fuel additives for control of deposits formation should be investigated.

5. Deposits found in operational aircraft should be analyzed.

WORKING GROUP V - FUEL SAFETY

Reduced Flashpoint Fuels

The group was in general agreement that some reduction in flashpoint below the present 100° F minimum was feasible from a safety standpoint; specifically, a level as low as 80° to 90° F was considered acceptable. The primary factors considered were

1. Effect of ground ambient temperature
2. Flame spreading rate over quiescent spills
3. Fueling and in-flight operations

Because of the general absence of ignition sources in flight, and the fact that wing tank design is predicated on the vapor space always being in the flammable range, this mode of aircraft operation was not expected to be affected by flashpoint reduction. Continued vigilance in wing tank design was recommended to prevent introduction of ignition sources, such as lightning and hot fragments from disintegrating engines.

Some increase in hazard was predicted during fueling operations particularly from static induced ignitions or from spills. The group determined that the static hazard could be controlled by an antistatic additive and that the fire danger from spills was no greater than that experienced with JP-4 and with Jet A in hot climates. In both these cases, experience was excellent.[1]

It was agreed that use of higher volatility fuels would, at least directionally, tend to reduce the potential benefits of antimisting additives. Any programs in this area should be extended to include fuels of intermediate flashpoint level (70° to 90° F).

Higher End Point Fuel

Some reduction in the in-flight restart envelope can be managed without operational problems. The magnitude is unknown, and it is recommended that NASA include this testing in any program involving a broadened distillation specification.

Laboratory-determined autoignition temperature (AIT) will be unaffected by end points up to 650° F, and there will be little or no effect on misting in severe crashes. However, there may be some reduction in minimum hot-surface ignition temperatures due to the presence of increased-molecular-weight fractions in the higher-boiling-point components of the fuel.

Regulatory Code Variations

The wide variance in regulatory codes by different municipalities, states, and countries will have a significant effect on the acceptance of reduced-flashpoint fuels. In the U.S., these problems will be minimal because aviation fuels are not regulated. Some overseas jurisdictions will just not accept a reduced-flashpoint jet fuel (at the present time).

[1]Several members of the panel were quite vociferous in their opinion that experience with JP-4 (from a safety standpoint) was excellent and that "closing the door" on this source of supply was unwarranted. This position was not unanimous.

The marketing practice of dual branding of kerosene for ground and aviation use will strongly influence the acceptance of a lower flashpoint product, since flashpoints below 100° F are unsafe for home use.

1975 Coordinating Research Council Fuel Safety Report

Based on information available at the time of the meeting, the use of higher volatility fuels has not been an adverse factor in any aircraft incidents since publication of the 1975 CRC report. The group sees no reason to invalidate the report's conclusions.

WORKING GROUP PARTICIPANTS

General Chairman:
    John P. Longwell   Massachusetts Institute of Technology

NASA Lewis Research Center Observers:
    Warner L. Stewart   Director of Aeronautics
    Richard A. Rudey   Chief, Airbreathing Engines Division
    Donald A. Petrash   Chief, Combustion and Pollution Research
                                         Branch

Working Group I - Aviation Fuels Supply and Demand:
    J. Morley English, Chairman   University of California, Los Angeles
    Jorgen Birkeland   ERDA Headquarters
    Paul P. Campbell   United Airlines
    William G. Dukek   Exxon Research and Engineering
    Herbert R. Lander   Air Force Aeropropulsion Laboratory
    John J. Madison   NASA Headquarters, Aircraft Energy Effi-
                                         ciency Office
    Larry Maggitti   Naval Air Propulsion Test Center
    John E. Mykytka   Pan American World Airways, Inc.
    A. J. Pasion   Boeing Commercial Airplane Co.
    Donald L. Rhynard   Mobil Research and Development Co.
    Roger W. Saari   Eastern Air Lines, Inc.
    Jeffrey L. Smith   Consultant, Econergy, Inc., Los Angeles
    Jack Grobman,
        NASA Lewis coordinator

Working Group II - Combustion Research and Technology for Broad-Specification Fuels:
    Donald W. Bahr, Chairman   General Electric Co., Aircraft Gas Turbine
                                         Division
    W. S. Blazowski   Exxon Research and Engineering
    Perry Goldberg   Pratt & Whitney Aircraft Group
    John M. Haasis   AiResearch Manufacturing of Arizona
    John Herrin   Pratt & Whitney, Government Products
                                       Division, Florida
    C. J. Jachimowski   NASA Langley Research Center
    Alan Moses   Army Fuels and Lubricants Research Lab-
                                       oratory

George Opdyke, Jr.	AVCO Corp., Lycoming Division
Thomas Rosfjord	Air Force Aeropropulsion Laboratory
Daniel E. Sokolowski	NASA Lewis Research Center
Louis Spadaccini	United Technologies Research Center
Gerald G. Tomlinson	General Motors, Detroit Diesel-Allison Division

Helmut F. Butze,
   NASA Lewis coordinator

Working Group III - Fuel System Research and Technology for Broad-Specification Fuels:

Ivor Thomas, Chairman	Boeing Commercial Airplane Co.
Robert T. Holmes	Shell Oil Co.
Charles R. Martell	Air Force Aeropropulsion Laboratory
David J. Miller	NASA Headquarters, Aeronautical Propulsion Division
Walter D. Sherwood	Trans World Airlines, Inc.
Kurt H. Strauss	Texaco, Inc.
Edward F. Versaw	Lockheed-California Co.

Robert Friedman,
   NASA Lewis coordinator

Working Group IV - Fuel Thermal Stability Research and Technology:

Jack A. Bert, Chairman	Chevron Research
Dennis W. Brinkman	ERDA Bartlesville Energy Research Center
Steven R. Daniel	Colorado School of Mines
Allyn R. Marsh, Jr.	Pratt & Whitney Aircraft Group
C. J. Nowack	Naval Air Propulsion Test Center
M. W. Shayeson	General Electric Co., Aircraft Gas Turbine Division
William F. Taylor	Exxon Research and Engineering
Alexander Vranos	United Technologies Research Center

Thaine W. Reynolds,
   NASA Lewis coordinator

Working Group V - Fuel Safety:

John H. Warren, Chairman	Mobil Sales and Supply Corp.
Ben Botteri	Air Force Aeropropulsion Laboratory
Robert N. Hazlett	Naval Research Laboratory
Robert R. Hibbard	NASA Lewis Research Center
Stanley Jones	Pan American World Airways, Inc.

William Maxwell   Mobil Research and Development
A. T. Peacock   Douglas Aircraft Co.
I. Irving Pinkel   Consultant, Fairview Park, Ohio
Solomon Weiss   NASA Lewis Research Center
Albert C. Antoine,
    NASA Lewis coordinator

1. Report No. NASA CP-2033	2 Government Accession No	3. Recipient's Catalog No.	
4. Title and Subtitle JET AIRCRAFT HYDROCARBON FUELS TECHNOLOGY		5. Report Date January 1978	
		6. Performing Organization Code	
7 Author(s) John P. Longwell, Editor Massachusetts Institute of Technology		8. Performing Organization Report No E-9457	
9. Performing Organization Name and Address		10. Work Unit No.	
		11. Contract or Grant No.	
12. Sponsoring Agency Name and Address National Aeronautics and Space Administration Washington, D.C. 20546		13. Type of Report and Period Covered Conference Publication	
		14. Sponsoring Agency Code	

15. Supplementary Notes

16. Abstract

In a workshop on jet aircraft hydrocarbon fuels technology, held at the Lewis Research Center, June 7-9, 1977, a broad-specification, referee fuel was proposed for research and development. This fuel has a lower, closely specified hydrogen content and higher final boiling point and freezing point than ASTM Jet A. The workshop recommended various priority items for fuel research and development. Key items include prediction of trade-offs among fuel refining, distribution, and aircraft operating costs; combustor liner temperature and emissions studies; and practical simulator investigations of the effect of high-freezing-point and low-thermal-stability fuels on aircraft fuel systems.

17. Key Words (Suggested by Author(s)) Fuel; Aviation fuels; Turbine fuels; Jet fuels; Hydrocarbon fuels; Combustor; Freezing point; Hydrogen content; Thermal stability; Flashpoint; Pumpability; Hydrotreating; Aircraft fuel system		18. Distribution Statement Unclassified - unlimited STAR Category 28	
19. Security Classif. (of this report) Unclassified	20 Security Classif. (of this page) Unclassified	21 No. of Pages 02	22. Price* A04